The Gastrointestinal Circulation

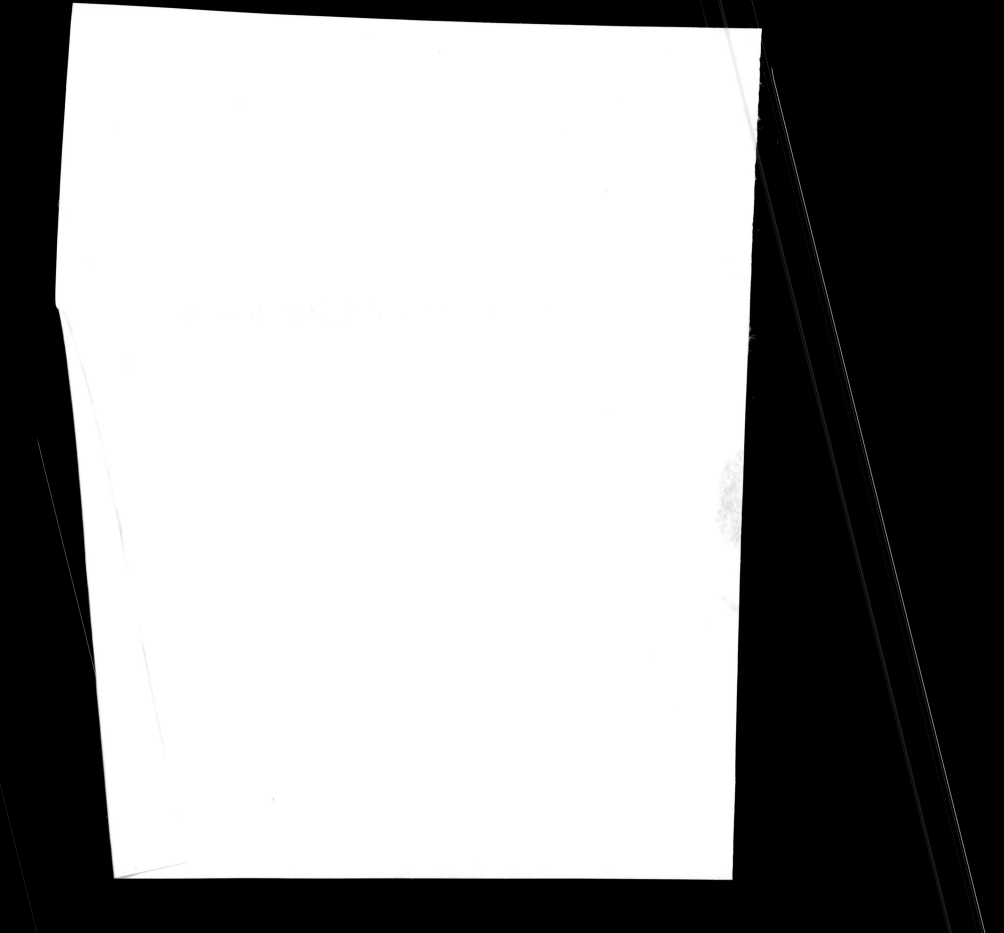

Integrated Systems Physiology: From Molecule to Function

Editors

D. Neil Granger, *Louisiana State University Health Sciences Center*

Joey P. Granger, *University of Mississippi Medical Center*

Physiology is a scientific discipline devoted to understanding the functions of the body. It addresses function at multiple levels, including molecular, cellular, organ, and system. An appreciation of the processes that occur at each level is necessary to understand function in health and the dysfunction associated with disease. Homeostasis and integration are fundamental principles of physiology that account for the relative constancy of organ processes and bodily function even in the face of substantial environmental changes. This constancy results from integrative, cooperative interactions of chemical and electrical signaling processes within and between cells, organs and systems. This eBook series on the broad field of physiology covers the major organ systems from an integrative perspective that addresses the molecular and cellular processes that contribute to homeostasis. Material on pathophysiology is also included throughout the eBooks. The state-of the-art treatises were produced by leading experts in the field of physiology. Each eBook includes stand-alone information and is intended to be of value to students, scientists, and clinicians in the biomedical sciences. Since physiological concepts are an ever-changing work-in-progress, each contributor will have the opportunity to make periodic updates of the covered material.

Published titles

(for future titles please see the website, www.morganclaypool.com/page/lifesci)

Capillary Fluid Exchange: Regulation, Functions, and Pathology

Joshua Scallan, Virgina H. Huxley, and Ronald J. Korthuis

2010

The Cerebral Circulation

Marilyn J. Cipolla

2009

Hepatic Circulation
W. Wayne Lautt
2009

Platelet-Vessel Wall Interactions in Hemostasis and Thrombosis
Rolando Rumbaut and Perumal Thiagarajan
2010

The Gastrointestinal Circulation
Peter R. Kvietys
www.morganclaypool.com

ISBN: 9781615041176 paperback

ISBN: 9781615041183 ebook

DOI: 10.4199/C00009ED1V01Y201002ISP005

A Publication in the Morgan & Claypool Life Sciences Publishers series

INTEGRATED SYSTEMS PHYSIOLOGY: FROM MOLECULE TO FUNCTION

Book #5

Series Editors: D. Neil Granger, Louisiana State University; Joey Granger, University of Mississippi

Series ISSN Pending

The Gastrointestinal Circulation

Peter R. Kvietys
London Health Sciences Center

INTEGRATED SYSTEMS PHYSIOLOGY: FROM MOLECULE TO FUNCTION #5

MORGAN&CLAYPOOL LIFE SCIENCES

ABSTRACT

The microcirculation of the gastrointestinal tract is under the control of both myogenic and metabolic regulatory systems. The myogenic mechanism contributes to basal vascular tone and the regulation of transmural pressure, while the metabolic mechanism is responsible for maintaining an appropriate balance between O_2 demand and O_2 delivery. In the postprandial state, hydrolytic products of food digestion elicit a hyperemia, which serves to meet the increased O_2 demand of nutrient assimilation. Metabolically linked factors (e.g., tissue pO_2, adenosine) are primarily responsible for this functional hyperemia. The fenestrated capillaries of the gastrointestinal mucosa are relatively permeable to small hydrolytic products of food digestion (e.g., glucose), yet restrict the transcapillary movement of larger molecules (e.g., albumin). This allows for the absorption of hydrolytic products of food digestion without compromising the oncotic pressure gradient governing transcapillary fluid movement and edema formation. The gastrointestinal microcirculation is also an important component of the mucosal defense system whose function is to prevent (and rapidly repair) inadvertent epithelial injury by potentially noxious constituents of chyme. Two pathological conditions in which the gastrointestinal circulation plays an important role are ischemia/reperfusion and chronic portal hypertension. Ischemia/reperfusion results in mucosal edema and disruption of the epithelium due, in part, to an inflammatory response (e.g., increase in capillary permeability to macromolecules and neutrophil infiltration). Chronic portal hypertension results in an increase in gastrointestinal blood flow due to an imbalance in vasodilator and vasoconstrictor influences on the microcirculation.

KEYWORDS

blood vessels, lymphatics, metabolic regulation, myogenic regulation, vascular permeability, postprandial hyperemia, mucosal defense, Starling forces, ischemia/reperfusion, portal hypertension

Contents

1. Introduction 1

2. Anatomy 3
    2.1 Extramural Blood and Lymphatic Vessels 3
    2.2 Intramural Blood and Lymphatic Vessels 3

3. Regulation of Vascular Tone and Oxygenation 9
    3.1 Basal Hemodynamics and Oxygenation 9
    3.2 Intrinsic Vasoregulation: Myogenic and Metabolic 10
         3.2.1 Myogenic 10
         3.2.2 Metabolic 13
    3.3 Relative Impact of Myogenic and Metabolic Mechanisms 17
         3.3.1 Reactive Hyperemia 17
         3.3.2 Venous Pressure Elevation 17
         3.3.3 Arterial Pressure Reduction 18
    3.4 Mediators of Metabolic Vasoregulation 19
         3.4.1 Tissue pO_2 19
         3.4.2 Adenosine 20
         3.4.3 Nitric Oxide 20
    3.5 Mediators of Myogenic Vasoregulation 21
    3.6 Shear Stress Modulation of Metabolic and Myogenic Regulatory Systems 21

4. Extrinsic Vasoregulation: Neural and Humoral 23
    4.1 Neural 23
         4.1.1 Postganglionic Sympathetic 23
         4.1.2 Sensory C Fibers 24
         4.1.3 Enteric Nerves 25
    4.2 Circulating Vasoactive Substances 26

5.   Postprandial Hyperemia 29
    5.1   General Characteristics 29
    5.2   Localization of the Postprandial Hyperemia 30
    5.3   Constituents of Chyme Responsible for the Postprandial Hyperemia 30
    5.4   Mechanisms Involved in the Postprandial Hyperemia 32
        5.4.1   Extrinsic Nerves 32
        5.4.2   Enteric Neural Reflexes 32
        5.4.3   Circulating Hormones 33
        5.4.4   Tissue Metabolic Activity 33

6.   Transcapillary Solute Exchange 37
    6.1   Ultrastructural Pathways 37
        6.1.1   Endothelial Cell Membrane 37
        6.1.2   Fenestrae 38
        6.1.3   Pinocytotic Vesicles 38
        6.1.4   Interendothelial Cell Junctions 38
        6.1.5   Glycocalyx 38
        6.1.6   Basement Membrane 39
    6.2   Physiological (Functional) Pathways 39
        6.2.1   Small Solutes 39
        6.2.2   Macromolecules 40
    6.3   Factors Influencing Vascular Permeability 47
    6.4   Ultrastructural Correlates for the Functional Pathways 49

7.   Transcapillary Fluid Exchange 53
    7.1   Net Transcapillary Fluid Movement ($J_{v,c}$) 53
    7.2   Capillary Filtration Coefficient ($K_{f,c}$) 54
    7.3   Capillary Pressure (P_c) 55
    7.4   Interstitial Fluid Pressure (P_t) 55
    7.5   Osmotic Reflection Coefficient (σ_d) 57
    7.6   Transcapillary Oncotic Pressure Gradient ($\pi_c - \pi_t$) 57

8.   Interaction of Capillary and Interstitial Forces 59
    8.1   Increased Venous Pressure 59
    8.2   Decreased Arterial Pressure 64
    8.3   Transepithelial Fluid Absorption 65

            8.3.1   Glucose and/or Electrolyte-Coupled Fluid Absorption 66
            8.3.2   Oleic Acid-Coupled Fluid Absorption................................... 68
            8.3.3   Colonic Fluid Absorption 70
    8.4   Transepithelial Fluid Secretion... 70

9.   **Gastrointestinal Circulation and Mucosal Defense** 73
    9.1   Gastrointestinal Acid Load 73
    9.2   Intestinal Lipid Load... 75
    9.3   Gastrointestinal Restitution.. 77

10.   **Gastrointestinal Circulation and Mucosal Pathology I: Ischemia/Reperfusion**..... 79
    10.1   Moderate Reductions in Blood Flow: Dysfunction........................... 79
    10.2   Severe Reductions in Blood Flow: Injury 80
            10.2.1   Ischemia-Induced Injury 80
            10.2.2   I/R-Induced Injury... 81
            10.2.3   I/R-Induced Inflammation..................................... 82
            10.2.4   Luminal Factors May Aggravate Mucosal Injury During I/R............. 85
    10.3   Ischemic Tolerance and Restitution 86

11.   **Gastrointestinal Circulation and Mucosal Pathology II:
    Chronic Portal Hypertension**... 87
    11.1   The Gastrointestinal Circulation in Chronic Portal Hypertension (PH)......... 87
    11.2   Collateral Vessels: Portosystemic Shunting 89
    11.3   Gastrointestinal Hyperemia... 90
    11.4   Transcapillary Fluid Exchange 91
    11.5   Luminal Factors May Aggravate Mucosal Injury
            During Portal Hypertension.. 91

12.   **Summary and Conclusions** 93

References 95

Author Biography 127

C H A P T E R   1

Introduction

The primary function of the gastrointestinal tract is the digestion and absorption of ingested food and water. The gastrointestinal blood circulation is responsible for delivery of the oxygen required for the absorptive and secretory functions of the mucosa and the motor activity of the muscularis. The distribution of absorbed nutrients and water to all the organs of the body is accomplished by way of both the blood and lymph circulations. At rest, the gastrointestinal tract accounts for one-fourth of total body oxygen consumption and cardiac output, and this demand is increased postprandially. Herein, the overall objective is to summarize the physiologic role of the blood and lymph vessels in supporting the overall absorptive, secretory, and motor function of the stomach, small intestine, and colon with an emphasis on the mucosal microcirculation. The major issues to be addressed include: (1) the vasoregulatory mechanisms involved in meeting the enhanced O_2 demand of assimilation of ingested nutrients and (2) the alterations in forces and membrane parameters governing transcapillary fluid exchange, which allow for efficient delivery of nutrients to other organs without compromising mucosal integrity. In addition, two pathologic conditions in which the gastrointestinal microcirculation plays a critical role will be addressed: ischemia/reperfusion (I/R) and portal hypertension (PH).

· · · ·

CHAPTER 2

Anatomy

2.1   EXTRAMURAL BLOOD AND LYMPHATIC VESSELS

The major arteries supplying the gastrointestinal tract are the celiac, superior mesenteric, and inferior mesenteric arteries. The celiac supplies the stomach and the proximal portion of the small intestine (duodenum), the superior mesenteric supplies the rest of the small intestine and proximal portion of the colon, while the inferior mesenteric supplies the distal portion of the colon. The areas supplied by these three major arterial conduits are not discrete, since there are numerous arcades of smaller arteries along the mesenteric border which anastomose with one another and provide collateral blood flow. These arcades give rise to vasa recta, whose branches encircle the musculature of the stomach, small intestine, and colon and, ultimately, penetrate the muscularis and form an arterial plexus within the submucosa [1–3].

The small veins draining the gastrointestinal tract generally parallel the arterial circuitry, including the anastomoses, and deliver the venous effluent to the portal vein via three major tributaries. The splenic vein drains the stomach, the superior mesenteric vein drains the upper small intestine, while the inferior mesenteric vein drains the distal portions of the colon. These three tributaries drain into the portal vein, which supplies the liver whose venous effluent is delivered back to the heart [2,4,5].

The lymph vessels draining the gastrointestinal tract run predominantly in association with blood vessels and enter various lymph nodes. The efferent lymphatic vessels from the lymph nodes empty into the cisterna chyli and join the systemic circulation via the thoracic duct [2,4–6].

2.2   INTRAMURAL BLOOD AND LYMPHATIC VESSELS

In general, the major arterial vessels supplying the mucosal and muscular layers of the gastrointestinal tract originate from the arterial plexus located in the submucosa. Arterioles from the submucosa branch into capillary networks in the mucosa and in the longitudinal and circular muscle layers where they run in parallel to the smooth muscle fibers. In general, the arterial supply of the mucosa and the muscularis layers of the gastrointestinal tract are arranged in parallel allowing for independent control of the blood supply to these two regions [7,8]. The lymphatic vessels draining the mucosa and the muscularis empty into the submucosal network of collecting lymphatics. In the muscularis, the lymphatic vessels run close to blood vessels with frequent anastomoses with each other [6].

The microcirculations of the mucosal and muscularis layers support important functional activities, such as absorption/secretion and motor activity, respectively. The mucosal layer receives approximately 80% of the total intramural blood flow; the muscularis receiving the remaining 20% [9–11]. This is presumably due to the more demanding metabolic activity of this layer. The mucosal microcirculation has a much more complex architecture than that of the muscularis, and there are some striking anatomical differences between the mucosal microcirculation of the small intestine and that of the stomach and colon.

In the stomach, submucosal arterioles branch into capillaries at the base of the glands and pass along the glands to the luminal surface of the mucosa where they form a luminal capillary net-

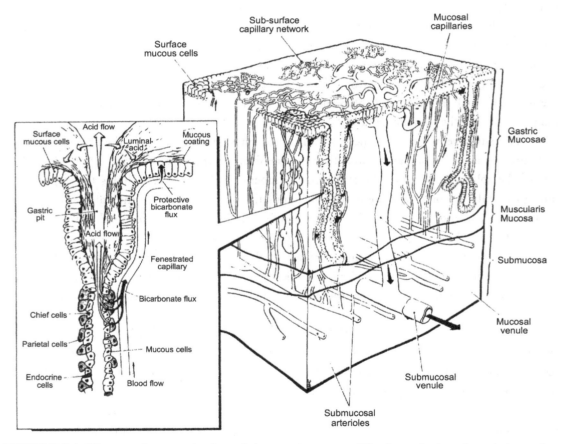

FIGURE 2.1: The vascular organization of the gastric mucosa. The inset depicts the microvascular transport of HCO_3^- from the acid secreting portion of the gastric pit to the surface epithelial cells. Used with permission from *Gastroenterology* 1984; pp. 866–875.

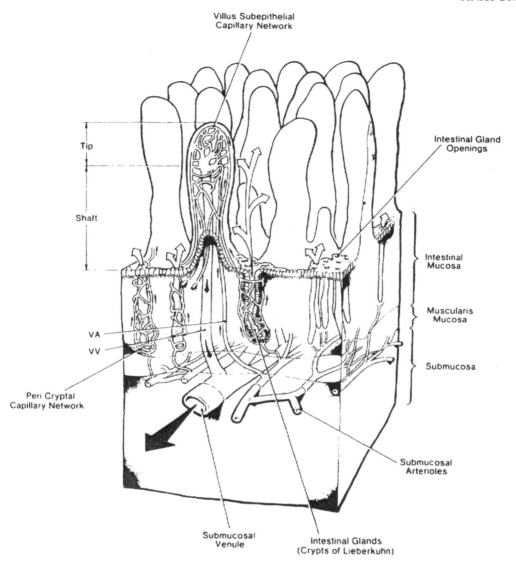

FIGURE 2.2: The vascular organization of the small intestinal mucosa. VA, villus arteriole; VV, villus venule. Used with permission from *Microvasc. Res.* 1972; 4: pp. 62–76.

work (Figure 2.1) [12]. The capillary network surrounding the glands is drained by venules near the luminal surface of the mucosa and pass directly to the submucosal venous plexus without receiving any direct capillary tributaries within the mucosa [8,12]. When viewed from the mucosal surface by confocal endomicroscopy, the capillary networks surrounding the glands of the gastric body exhibit a honeycomb-like appearance [13]. The gastric pits of the antrum are surrounded by a "coil-shaped"

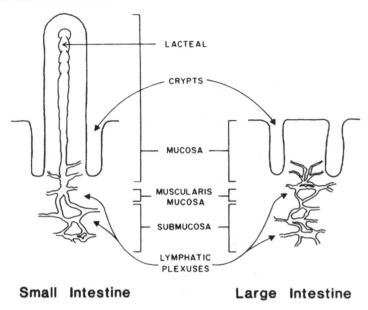

Small Intestine   **Large Intestine**

FIGURE 2.3: The mucosal–submucosal lymphatic organization of the small and large intestine. Used with permission from *Gastroenterology* 1981; 81: pp. 1080–1090.

capillary network. The initial lymphatics are located below the gastric glands as a plexus. In general, no lymph vessels are found in the upper portion of the gastric mucosa [5,6,14].

In the small intestine, the submucosal arterioles enter the mucosa to form the villus microcirculation whose pattern varies among species [5]. In general, within human villi, there is an eccentrically located single arteriole, which passes to the tip and forms a capillary fountain or tuft-like network with numerous anastomoses with the single eccentrically located venule (Figure 2.2) [15]. The villus capillaries are situated within 2 μm of the epithelial cells [16]. The crypt capillary network (also derived from submucosal arterioles) supplies the shaft of the villus and also drains into the venule exiting the villus.

The lymphatic system of the small intestine originates as a large centrally located vessel (lacteal) within the villi (Figure 2.3). The apical portion of the lacteal has a "cul-de-sac" endothelium, ensuring the propulsion of lymph toward the collecting lymphatics when the villus contracts. Although similar in size to the venular capillaries, they lack endothelial cell junctions, presumably to aid in the transport of chylomicra [5,17].

The colonic mucosal microvascular arrangement is similar to that of the stomach. The feeding arterioles and their capillary branches pass along the glands to the luminal surface of the mucosa where they form a capillary network surrounding the glands, presenting a honeycomb appearance when viewed from the surface [18,19]. The capillary density within the honeycomb networks is greater in the proximal colon than in the distal portion [20]. The colonic capillaries are situated much closer to the epithelium (1 μm) than their counterparts in the small intestine [16]. The initial lymphatics of the colon originate near the basal aspect of the gastric glands where they form plexi (Figure 2.3); the upper portion of the colonic mucosa is devoid of lymphatic vessels [16,21].

CHAPTER 3

Regulation of Vascular Tone and Oxygenation

3.1 BASAL HEMODYNAMICS AND OXYGENATION

Estimates of blood flow (in ml/min × 100 g) to the quiescent gastrointestinal tract of dogs and cats range from 15 to 100 for the stomach, 35–120 for the small intestine, and 10–74 for the colon [7,22–24]. Somewhat lower values for resting blood flow to the stomach (11), small intestine (29–70), and colon (8–35) have been reported for man [7]. Although reported resting values vary between and within a given species, it is generally agreed that small intestinal blood flow exceeds that of the stomach or colon. Further, as mentioned above, throughout the gastrointestinal tract, the mucosal layer receives a greater portion of the total wall flow than the muscularis region.

The delivery of oxygen and nutrients to the gastrointestinal parenchyma is determined not only by blood flow but also by the number of capillaries opened to blood perfusion (capillary density). It is generally assumed that approximately 25–50% of the capillaries in the quiescent gastrointestinal tract are open to perfusion [25]. Experimentally, the capillary filtration coefficient ($K_{f,c}$) can provide an index of capillary density or functional exchange capacity in a tissue [26,27]. $K_{f,c}$ has been assessed in the stomach, small intestine, and colon using volumetric or gravimetric techniques. These approaches involve a rapid increase in venous pressure, while measuring changes in organ volume or weight. The slope of the volume (or weight) increase (attributed to capillary filtration) is divided by the increment in capillary pressure to yield $K_{f,c}$. One caveat to this approach is that $K_{f,c}$ is a measure of transcapillary hydraulic conductance and, as such, is influenced by both the capillary surface area available for exchange as well as the capillary permeability to solutes and fluid [26,27]. In general, however, unless the experimental maneuvers involve changes in permeability (e.g., injury, inflammation, agents that affect endothelial integrity), changes in $K_{f,c}$ are taken as changes in perfused capillary density or exchange capacity.

The resting values of $K_{f,c}$ range from 0.03 to 0.30 ml/min/100 g for the small intestine of cats, dogs, and rats [28–30]. A similar range of $K_{f,c}$ values have been reported for the dog stomach and colon [31–34]. The relative uniformity of the ranges of the mean values for $K_{f,c}$ among the different regions of the gastrointestinal tract preclude any definitive statements regarding their relative capillary surface areas.

Oxygen consumption or demand of gastrointestinal organs can be calculated as the product of blood flow (Q_B) and the arteriovenous oxygen difference ($(A - V)O_2$). Resting values of oxygen consumption of the stomach, small intestine, and colon in the dog and cat are fairly uniform; ranging from 1.5 to 2.5 ml/min × 100 g [35–40]. There are no apparent differences in the resting values for oxygen consumption among the various regions of the gastrointestinal tract. Oxygen consumption of the rat small intestine is about 4.8 ml/min × 100 g [29], values two to three times higher than those reported for larger animals.

3.2   INTRINSIC VASOREGULATION: MYOGENIC AND METABOLIC

Gastrointestinal blood flow and oxygenation are maintained within relatively narrow limits by vasoregulatory mechanisms operating at the organ/tissue level. Existence of intrinsic regulatory systems has been established by experiments in isolated organs or blood vessels. Data obtained from experimental studies have been incorporated into mathematical models allowing for predictions of microcirculatory behavior not amenable to experimentation. Based on both experimental and computer modeling approaches, two major mechanisms have been invoked to explain the ability of the gastrointestinal tract to regulate its vasculature to meet tissue demands: the myogenic and metabolic (Figure 3.1) [7,25,41–49].

3.2.1   Myogenic

Isolated mesenteric arterioles respond to elevations in transmural pressure with vasoconstriction and vice versa; the response being the most intense in the smaller resistance vessel [50]. This phenomena is referred to as the myogenic response [51–53]. The myogenic response is an intrinsic property of vascular smooth muscle, since it occurs in mesenteric arterioles that have had their endothelial lining removed [50]. The myogenic response is based on the law of LaPlace, which contends that tension development in smooth muscle is the product of the intravascular pressure and vessel radius (Figure 3.1). Thus, increases in intravascular pressure lead to decreases in radius (constriction) in an attempt to maintain tension. Conversely, decreases in vascular transmural pressure result in dilation to maintain tension.

If a myogenic mechanism is operative in the microcirculation of the gastrointestinal tract, an increase in microvascular transmural pressure would result in increases in vascular resistance and vice versa (Figure 3.1). Since as much as 70% of the increase in venous pressure can be transmitted back upstream to the capillaries and resistance vessels, a commonly used experimental maneuver is to increase venous pressure in isolated preparations of the stomach, small intestine, or colon. To determine whether resistance vessels are affected by acute venous hypertension, arterial and venous

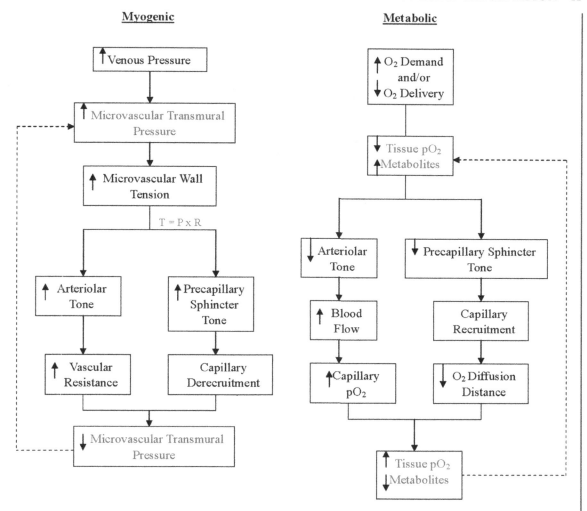

FIGURE 3.1: Myogenic and metabolic theories of intrinsic regulation of the microcirculation in the gastrointestinal tract. Modified from and used with permission from *Handbook of Physiology, The Gastrointestinal System I*, Chapter 39, 1989, pp. 1405–1474.

pressures (arteriovenous pressure gradient; ΔP) as well as blood flow (Q_B) are measured and resistance calculated ($R = \Delta P / Q_B$). To assess the myogenic impact on exchange vessels, the capillary filtration coefficient ($K_{f,c}$) can be measured. The existence of a myogenic regulatory system has been experimentally verified in the stomach [24], small intestine [27,54,55], and colon [40,56,57].

Almost invariably, there is an increase in vascular resistance (and decrease in blood flow) when venous pressure is elevated (Figure 3.2). In addition to the regulation of resistance vessels, the

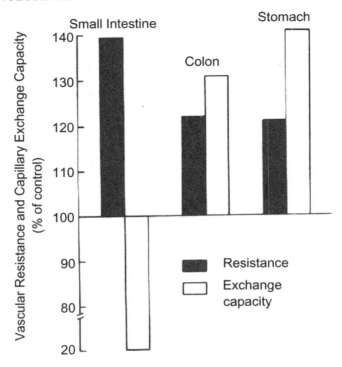

FIGURE 3.2: Effects of venous pressure elevation (from 0 to 20 mmHg) on vascular resistance and capillary exchange capacity (capillary filtration coefficient) in the gastrointestinal tract. Modified from and used with permission from *Pathophysiology of the Splanchnic Circulation*, Vol I, Chapter 1, 1987, pp. 1–56.

myogenic response may play a role in regulating precapillary sphincter tone and thereby capillary exchange capacity. There is evidence to support an increase in precapillary resistance (decrease in $K_{f,c}$) during acute venous hypertension [28]. However, elevations in microvascular transmural pressure increase, rather than decrease, capillary exchange capacity in the stomach and the colon (Figure 3.2) [41,42]. The latter responses are rather inconsistent with the generally held view that smaller blood vessels are more sensitive to the alterations in transmural pressure then larger vessels, i.e., the smaller vessels exhibit greater myogenicity [52,53].

The physiologically significant roles attributed to intrinsic myogenicity of blood vessels in the gastrointestinal tract are (1) establishment and maintenance of basal vascular tone and (2) regulation of blood flow during acute changes in perfusion pressure [52,58]. An extension of the latter role of the myogenic response may be to dampen the arterial pulse pressure at the level of the microcirculation thereby ensuring a more uniform flow distribution to the capillaries [59]. It is noteworthy that while the myogenic mechanism plays an important role in maintaining capillary pressure (and glomerular filtration rate) in the kidney [51], autoregulation of capillary pressure (and capillary filtration rate) in the small intestine is rather poor [60].

3.2.2   Metabolic

The metabolic theory of local blood flow holds that tissue oxygenation, specifically tissue pO_2, is the regulated variable (Figure 3.1). According to the metabolic theory, an increase in metabolic activity (increased O_2 demand) would lead to an increased (1) consumption of O_2 by the mitochondria resulting in a decrease in tissue pO_2 and (2) generation of vasodilator by-products of metabolism. The subsequent increases in blood flow (resistance vessels) and capillary surface area (precapillary sphincters) would tend to increase tissue pO_2 and washout out vasodilator metabolites, thereby closing the feedback loop [25,42,43,61,62].

Both arteriolar resistance and precapillary sphincters play a role in the regulation of tissue oxygenation. Resistance vessels regulate the convective delivery of arterial O_2 to capillaries by modulating blood flow. Thus, the rate of oxygen delivery to capillaries is the product of blood flow and arterial O_2 concentration. Precapillary sphincters regulate the diffusive delivery of oxygen to cells/tissues by modulating the number of perfused capillaries. The number of perfused capillaries determines the effective capillary surface area available for oxygen diffusion as well as the diffusional distance for O_2 from capillaries to cells. If a metabolic mechanism is operative in the microcirculation of the gastrointestinal tract, an increase in the tissue O_2 demand-to-delivery ratio would result in decreases in vascular resistance and recruitment of capillaries (Figure 3.1).

The relative roles of resistance vessels and precapillary sphincters in maintaining tissue oxygenation during alterations in O_2 demand or O_2 delivery have been assessed by measurements of relevant parameters in isolated perfused preparations of the stomach, small intestine, or colon. Changes in resistance vessel tone can be assessed by measuring perfusion pressure and organ blood flow (Q_B). The capillary filtration coefficient ($K_{f,c}$) provides an estimate of capillary surface area. Despite the potential influence of vascular permeability on $K_{f,c}$, changes in $K_{f,c}$ under normal conditions (e.g., noninflammatory) is generally attributed to alterations in the number of perfused capillaries. In support of this contention is the observation that there is a direct linear correlation between $K_{f,c}$ and oxygen extraction [36,63]. O_2 consumption (or demand) can be calculated as the product of Q_B and the arteriovenous O_2 difference [$(A - V)O_2$]. The O_2 delivery-to-demand ratio can be readily calculated as the quotient of O_2 demand and O_2 delivery. The existence of a metabolic regulatory system has been experimentally verified in the stomach [35,64,65], small intestine [36,44,60,63,66], and colon [40,56,67]. A synopsis of these studies is presented below using the small intestine as a prototype of the gastrointestinal tract.

A common experimental maneuver to alter the O_2 delivery-to-demand ratio is to alter gastrointestinal blood flow (or O_2 delivery) either mechanically (pump-perfused preparations) or by graded stenosis of the arterial supply (naturally perfused preparations). In naturally perfused preparations, moderate decreases in perfusion pressure are associated with dilation of resistance vessels in an attempt to maintain O_2 delivery, and the capillary exchange capacity increases to enhance O_2 diffusion to cells. With substantial decreases in perfusion pressure, the decrease in vascular resistance is insufficient to maintain a constant organ blood flow, i.e., blood flow tends to decrease. Despite the fall in blood flow, oxygen consumption tends to remain at control levels due to increases in capillary exchange capacity ($K_{f,c}$) and an increase O_2 extraction by the tissue (Figure 3.3). Similar results have been noted in pump-perfused preparations with the added benefit of demonstrating that increases in blood flow (or O_2 delivery) are associated with decreases in O_2 extraction such that O_2 uptake is unaffected (Figure 3.3).

Metabolic regulatory mechanisms become exhausted with severe reductions in blood flow (O_2 delivery), and as a consequence, O_2 consumption (or O_2 demand) begins to decrease (Figure 3.3). At this point, oxygen consumption is directly dependent on blood flow (O_2 delivery). The resting blood flow of the gastrointestinal tract is greater than the critical level at which oxygen uptake becomes blood flow dependent. The reduction in oxygen uptake below a critical blood flow has been attributed to the effects of cellular pO_2 on mitochondrial O_2 consumption. The relationship between cell pO_2 and mitochondrial O_2 consumption is independent of cell pO_2 until a critically low cell pO_2 is achieved at which point mitochondrial O_2 uptake (consumption) is compromised. It is generally believed that resting cell pO_2 in the gastrointestinal tract is above the critical cell pO_2 [25]. For example, modest reductions in blood flow to the stomach reduce gastric epithelial cell pO_2 without compromising whole organ oxygen consumption [68]. However, when metabolic control of the microvasculature is inadequate to maintain capillary and tissue pO_2 above a critical level, diffusion of O_2 to cells is no longer adequate to maintain normal mitochondrial O_2 metabolism.

Another approach to alter the O_2 delivery-to-demand ratio is to simultaneously alter gastrointestinal O_2 demand and delivery (Figure 3.4) [63,69]. Increases in O_2 demand (enhanced secretory, absorptive, or motor activity) simply shift the plateau of the delivery (blood flow) to demand relationship upward [35,38,63], while decreases in the O_2 demand (temperature reduction) shift the plateau downward [63]. Collectively, these responses are consistent with the metabolic theory, i.e., tissue oxygenation, rather than blood flow per se, is the regulated variable.

Although in the experimental situation one can readily alter blood flow or the oxygen carrying capacity of the blood, it is virtually impossible to control capillary density. Indirect approaches are used to assess the role of exchange vessels in meeting the oxygen demands of the gut. For example, in constant flow preparations, increases in O_2 demand (nutrient absorption) or decreases in O_2 delivery (hypoxia) are associated with increases in capillary density as estimated by $K_{f,c}$ or [86]Rb

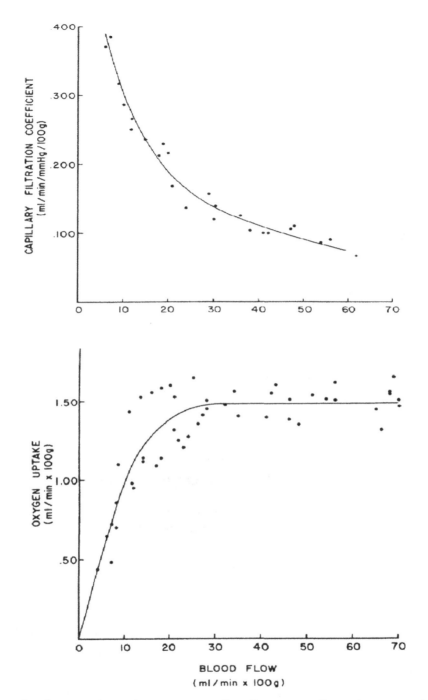

FIGURE 3.3: Steady-state relationships among capillary filtration coefficient (upper panel) and oxygen uptake (lower panel) and blood flow. Blood flow was altered by either reducing local arterial pressure (naturally perfused preparations) or by increasing blood flow with a pump interposed in the arterial circuit (pump-perfused preparations). Control resting blood flow was 37.4 ml/min × 100 g. Used with permission from *Am. J. Physiol.* 1982; pp. G570–G574.

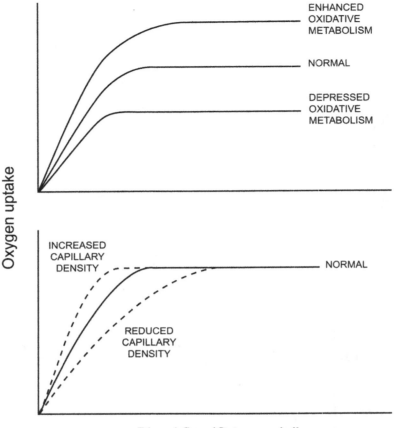

FIGURE 3.4: Relationship between oxygen uptake and oxygen delivery (blood flow) under normal conditions and during enhanced or depressed oxidative metabolism (upper panel). Relationship between oxygen uptake and oxygen delivery (blood flow) under normal conditions and during increased or reduced capillary density (lower panel). Used with permission from *Textbook of Gastroenterology*, Vol 1, Chapter 22, 2003, pp. 498–520.

extraction [70,71]. Based on mathematical modeling approaches [48], the expected influence of capillary density on the relationship between O_2 demand and O_2 delivery is depicted in Figure 3.4. If capillary density is increased then a greater reduction in blood flow or O_2 delivery would be tolerated before O_2 consumption was compromised; i.e., the curve would be shifted to the left.

The most significant physiological role attributed to metabolic regulation of blood vessels is the maintenance of an optimal tissue oxygenation during alterations in the O_2 demand-to-delivery ratio.

3.3   RELATIVE IMPACT OF MYOGENIC AND METABOLIC MECHANISMS

A variety of experimental perturbations have been used to assess the relative roles of the metabolic and myogenic mechanisms in the regulation of the gastrointestinal microcirculation. These include reactive hyperemia, vascular responses to venous pressure elevations, and pressure-flow autoregulation. It is becoming increasingly apparent that the myogenic and metabolic mechanisms may operate synergistically or in opposition to elicit a particular vascular response.

3.3.1   Reactive Hyperemia

The term "reactive hyperemia" refers to the transient increase in flow above basal levels after the release of an arterial occlusion. All of the regions of the gastrointestinal tract exhibit a reactive hyperemia after brief periods of arterial occlusion; the magnitude and duration of the hyperemia are directly related to the duration of the occlusion [42]. Both myogenic and metabolic hypotheses predict vasodilation upon release of an arterial occlusion; the myogenic based on the occlusion-induced fall in arteriolar transmural pressure and the metabolic based on the fall in tissue pO_2 and/or increase in vasodilator metabolites. However, only the metabolic mechanism can account for the direct relationship between the duration of occlusion and the magnitude of the hyperemia.

The reactive hyperemic response in the small intestine appears to be limited to the mucosal layer of the bowel wall [72]. The magnitude of the reactive hyperemia in the mucosa was related to the duration of arterial occlusion. This indicates that the microvasculature of the more metabolically active mucosa is more sensitive to metabolic factors. Further support for this contention is the observation that increasing the metabolic rate of the small intestine (mucosal nutrient transport) increases the magnitude of the reactive hyperemic response; which was again confined to the mucosal layer of the gut [72].

3.3.2   Venous Pressure Elevation

The myogenic and metabolic hypotheses predict opposite microvascular adjustments to increases in venous pressure. The myogenic hypothesis predicts that arteriolar resistance should increase, and capillary density should decrease in response to elevations in venous pressure due to the rise in intravascular pressure at the arteriolar and precapillary sphincter levels. By contrast, the metabolic hypothesis predicts that the elevation of venous pressure should decrease arteriolar resistance and increase capillary density due to the reduced blood flow and accumulation of vasodilator metabolites and/or decrease in tissue pO_2. In the stomach, small intestine, and colon, acute venous hypertension results in an increase in arteriolar resistance and decrease in organ blood flow (Figure 3.2), findings consistent with a myogenic mechanism. Isolated segments of mesenteric arteries also exhibit a

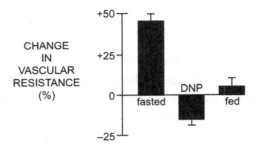

FIGURE 3.5: The effects of increasing intestinal oxygen demand by dinitrophenol (DNP) or instillation of digested food in the lumen (fed) on the vascular resistance response to venous pressure elevation. Used with permission from *Am. J. Physiol.* 1980; 238: pp. H836–H843.

myogenic vasoconstriction in response to increases in transmural pressure [73]. The response of the more sensitive microvascular elements (precapillary sphincters) to acute venous hypertension varies with the segment of the gastrointestinal examined. An increase in venous pressure results in a decrease in capillary exchange capacity in the small intestine, again, consistent with a myogenic mechanism. However, oxygen uptake by the small intestine is not compromised [44,61]. By contrast, acute venous hypertension in the stomach and colon results in an increase in capillary density (Figure 3.2). These latter findings indicate that the precapillary sphincters in the colon and stomach are more sensitive to metabolic factors, than those of the small intestine. Collectively, the experimental data indicate that both the myogenic and metabolic mechanisms are acting in concert in the quiescent gastrointestinal tract.

The microvascular response to acute venous hypertension is affected by the preexisting metabolic demand. When the O_2 demand of the small intestine and colon is increased, the characteristic increase in vascular resistance to acute venous hypertension is abolished [44,56]. For example, as shown in Figure 3.5, local administration of dinitrophenol or placement of digested food in the lumen prevents the increase in vascular resistance noted after acute venous hypertension. These findings indicate that when the small and large intestine are metabolically stressed, metabolic control overrides myogenic control of the local microcirculation.

3.3.3 Arterial Pressure Reduction

The ability of an organ to maintain a relatively constant blood flow in the face of fluctuations in arterial pressure is termed "pressure-flow autoregulation." Pressure-flow autoregulation has been demonstrated in the stomach [65], small intestine [29,60,74–76], and colon [40,56,57,77]. Both the myogenic and metabolic theories predict a similar vascular response during reductions in arterial pressure. According to both theories, a decrease in perfusion pressure should decrease arteriolar and precapillary sphincter resistance. The myogenic response would be elicited due to decreases in transmural pressure invoked by the local hypotension. The metabolic response would be elicited due to the initial local decrease in blood flow (O_2 delivery) resulting in a decrease in tissue pO_2 and/or build up of vasodilator metabolites. However, based on the following lines of evidence, it is generally believed that the metabolic mechanism prevails over the myogenic to elicit autoregulation. First, the autoregulatory ability of the more active mucosal layer of the stomach [78], small intestine [79], and colon [57] exceeds that of the whole organ. Second, when the metabolic demand of the small intestine [76,80] or colon [56] is increased by feeding or luminal instillation of transportable solutes, their autoregulatory ability is enhanced. Third, while autoregulation of blood flow is generally not perfect at lower perfusion pressures, oxygen uptake remains within normal limits [36,40,81]. The maintenance of tissue oxygen uptake despite the fall in blood flow is due to increases in capillary density [36,40,63].

In summary, both the myogenic and metabolic mechanisms play a role in intrinsic regulation of the gastrointestinal circulation. The myogenic mechanism appears to play a role in basal vascular tone and the regulation of transmural pressure in resistance vessels. The metabolic mechanism serves to maintain tissue oxygenation by regulating oxygen delivery to meet changes in oxygen demand. Unlike the case in the renal vasculature where myogenic regulation of capillary transmural pressure is critical [51], the myogenic mechanism appears to be readily overridden by metabolic factors in the gastrointestinal tract.

3.4   MEDIATORS OF METABOLIC VASOREGULATION

Of the various vasoactive mediators that have been proposed to be involved in mediating intrinsic metabolic regulation of the gastrointestinal microcirculation, tissue pO_2 and vasodilator metabolites (specifically, adenosine) have received the most attention [42,61,62]. In addition, nitric oxide (NO) has been implicated as a major mediator of metabolic vasoregulation. In general, neither tissue pO_2, adenosine, nor NO meets all of the criteria for being the decisive mediator of metabolic vasoregulation. More than likely, all three of them may act in concert or opposition depending on the prevailing situation. Evidence in favor of and against their participation in metabolic regulation of the gastrointestinal tract microcirculation is presented below.

3.4.1   Tissue pO_2

Mathematical models predict that tissue pO_2 may be a direct link between metabolic activity of the gastrointestinal tract and microvascular tone [25,82]. Direct measurements of tissue pO_2 (microelectrodes) indicate that there is an inverse correlation between villus pO_2 and the rate of blood flow through submucosal vessels supplying the villus [83,84]. Further, small intestinal oxygen demand (glucose absorption) reduces villus pO_2 and increases blood flow through submucosal vessels supplying the villus. However, blood flow to the muscularis layer of the small intestine also increases despite an unaltered tissue pO_2 in the vicinity. Further, maintenance of mesenteric pO_2 by alteration of ambient oxygen did not effect pressure-flow autoregulation in mesenteric blood vessels [85]. Although the mesentery is primarily connective tissue (little metabolic activity), collectively, these latter observations do suggest that other factors, besides tissue pO_2, may be involved in intestinal vasoregulation.

3.4.2   Adenosine

Metabolic by-products may also provide an indirect link between metabolic activity of the gastrointestinal tract and microvascular tone. With respect to potential vasoactive metabolites, adenosine, a potent intestinal vasodilator [86], has received the most attention. Intestinal interstitial adenosine levels are increased during reactive hyperemia [87] and increased metabolic demand of nutrient transport [88] as reflected by increased venous and lymphatic adenosine concentrations. Adenosine blockade decreases the reactive hyperemic response [89] and either decreases or has no effect on the hyperemia associated with luminal nutrients [89,90]. Similarly, adenosine blockade either decreases [91] or does not affect [89] intestinal pressure-flow autoregulation. This ambiguity has an apparent resolution in the observations that adenosine blockade consistently diminishes the reactive hyperemic response and pressure-flow autoregulation in the hypermetabolic (nutrient absorption) small intestine [89,92]. It has been proposed that while adenosine does not play a role in vasoregulation under resting conditions or during moderate changes in the O_2 delivery-to-demand ratio, it has a significant contribution to the vasodilation induced by more severe stresses [89].

One disconcerting aspect regarding the potential role of adenosine in intestinal vasoregulation is that upon intra-arterial infusion, it appears to (1) redistribute blood flow from the mucosa to the muscularis and (2) decrease capillary exchange capacity and tissue oxygen consumption [93,94]. It has been proposed that adenosine is a more potent vasodilator of the muscular layer than the mucosal layer, and during intra-arterial infusion, adenosine induces a "vascular steal" response redirecting blood flow to the muscularis which is less (1) metabolically active and (2) vascularized. The situation may be quite different during local compartmentalized release of adenosine during nutrient transport by the mucosa or contractions of the muscularis. For example, during nutrient transport, mucosal adenosine levels would be increased locally and produce a hyperemia confined to the mucosal layer; blood flow to the muscularis should remain unaltered.

3.4.3   Nitric Oxide

Another potential mediator of intrinsic vasoregulation in the gastrointestinal tract is NO released from endothelial cells [61,62]. NO is synthesized from L-arginine by endothelial nitric oxide synthase (eNOS) [95,96]. The NO generated by the endothelial cells diffuses to the vascular smooth muscle, where it induces vascular smooth muscle contraction via cGMP-dependent protein kinase signaling pathway resulting in decreases in intracellular Ca^{++} concentration.

Based on the use of inhibitors of eNOS (e.g., L-NAME), endothelial-derived NO appears to play a role in basal vascular tone in the stomach [97] and small intestine [98]. A major stimulus for NO release by endothelial cell eNOS is vessel wall shear stress (or blood flow); [99] the resultant flow-mediated vasodilation is more prevalent in resistance arterioles than precapillary sphincters [99,100]. Thus, it would be predicted that decreases in blood flow (and O_2 delivery) would inhibit shear stress-mediated NO production and result in vasoconstriction. By contrast, hypoxia can result in endothelial NO production independent of shear stress and result in an NO-mediated dilation of isolated blood vessels [101]. *In vivo*, inhibition of NOS (e.g., L-NAME) potentiates pressure-flow autoregulation, indicating that NO is not mediating pressure-flow autoregulation, but actually opposing it [102]. Thus, during decreases in O_2 delivery, NO does not seem to play a role in the metabolically mediated vasodilation.

However, there is some convincing evidence that NO may play a role in the hyperemia associated with increases in O_2 demand. The hyperemia associated with small intestinal absorption of glucose [103–105] or gastric acid secretion [97,106] is associated with local generation of NO and can be attenuated by blockade of NO bioactivity.

3.5   MEDIATORS OF MYOGENIC VASOREGULATION

As mentioned above, the myogenic response is an intrinsic property of vascular smooth muscle, since it occurs in mesenteric arterioles that have had their endothelial lining removed [50]. In this context, vascular smooth muscle is generally believed to be both the sensing (tension sensor) and effector (contractile response) element. The signaling mechanisms, which transduce changes in transmural pressure into smooth muscle contraction, are largely unknown. A model of mechano-sensing has been proposed, which includes roles for extracellular matrix/integrins, smooth muscle cytoskeleton, and mechanosensitive enzyme systems, transporters, and channels. The current status of this area has been recently reviewed [51].

However, the myogenic response in intact blood vessels *in vivo* may be more complex. For example, endothelial-derived NO may modulate the myogenic-induced vasoconstriction. An increase in stretch of blood vessels results in endothelial production of NO *in situ* [107] as does increased endothelial cell deformation or cyclic strain *in vitro* [108,109]. In this case, NO would be serving to antagonize the smooth muscle myogenic vasoconstriction.

3.6   SHEAR STRESS MODULATION OF METABOLIC
AND MYOGENIC REGULATORY SYSTEMS

It is becoming apparent that endothelial wall shear stress imposed by the flowing blood can modulate microvascular tone. For example, an increase in shear stress results in a dilation of resistance vessels, which results in returning wall shear stress toward original levels [110]. It is generally believed that this vasoregulatory phenomenon is mediated via the release of vasodilator substances from the endothelium, such as prostaglandins and NO [111]. Endothelial-derived NO appears to be more important than prostaglandins in flow-induced vasodilation in isolated mesenteric arteries [112].

The precise mechanisms by which endothelial cells transduce shear stress into adjacent smooth muscle contraction are not entirely clear, but the endothelial glycocalyx, extracellular matrix components, integrins, and cell junctional molecules may be involved [99]. For example, NO production and flow-dependent vasodilation of mesenteric arteries is inhibited by enzymatic degradation of the endothelial glycocalyx [113,114]. Evidence is also accumulating to indicate that platelet-endothelial cell adhesion molecule-1 (PECAM-1), an endothelial cell junctional molecule, may be a key endothelial cell mechanoresponsive molecule. First, mechanical "tugging" of endothelial cell PECAM-1 with magnetic beads results in activation (phosphorylation) of PECAM-1 [115]. Second, shear stressed-induced phosphorylation of PECAM-1 enhanced PECAM-1 association with and phosphorylation of eNOS as well as NO production, events not noted in PECAM-1-deficient endothelial cells [116]. Finally, flow-induced vasodilation is significantly blunted in isolated arterioles from PECAM-1-deficient mice [117].

Since both metabolic and myogenic regulatory factors modulate blood flow, it is possible that concomitant alterations in shear stress may modify the overall microvascular response. In the case of the myogenic response, endothelial-derived factors do not appear to play a role, since denudation of the endothelium has no affect on the myogenic constriction of mesenteric vessels induced by an increase in intraluminal pressure [50]. However, *in vivo* increases in perfusion pressure and blood flow would result in a situation where flow-mediated vasodilation would oppose myogenic-mediated vasoconstriction [113]. In the case of the metabolic-mediated vasoregulation, endothelial-derived NO may either antagonize or potentiate the vascular responses to perturbations in O_2 demand or delivery. For example, the intestinal hyperemia induced by an increase in O_2 demand (e.g., nutrient transport) may be initiated by an NO-independent mechanism (e.g., tissue pO_2, adenosine) and result in shear stress-induced, NO-dependent potentiation/prolongation of the hyperemia. By contrast, the intestinal vasodilation induced by reductions in O_2 delivery (e.g., decreases in blood flow) may be antagonized by NO [102]. Interestingly, mathematical modeling approaches based on experimental data predict that the combined influence of metabolic and myogenic mechanisms override the effects of shear stress in blood flow autoregulation [46].

·   ·   ·   ·

CHAPTER 4

Extrinsic Vasoregulation: Neural and Humoral

4.1 NEURAL
4.1.1 Postganglionic Sympathetic

Neural regulation of small intestinal and colonic blood flow is generally achieved at the level of the submucosal arterioles, which are predominantly innervated by sympathetic fibers emanating from the celiac, superior mesenteric, and inferior mesenteric ganglia (Figure 4.1); parasympathetic innervation is negligible [118,119]. Sympathetic neural input is rather complex consisting of both divergent and convergent pathways, but in general, different fibers innervate specific portions of the microcirculation, i.e., arterioles and venules. Activation of sympathetic nerves elicits vasoconstriction of small mesenteric arteries primarily due to release of ATP and norepinephrine [120]. ATP appears to be more important with low-frequency stimulation of the nerves, while norepinephrine plays a greater role with high-frequency stimulation. By contrast, activation of sympathetic nerves induces vasoconstriction of submucosal arterioles primarily by release of ATP; a role for norepinephrine, if any, is in presynaptic modulation of neurotransmitter release [118].

During prolonged stimulation of the postganglionic sympathetic nerves, the resistance vessels begin to dilate with the result that gastrointestinal blood flow returns toward control levels despite continued nerve stimulation, a phenomenon termed "autoregulatory escape [121,122]." Autoregulatory escape has also been noted in mesenteric resistance vessels *in vivo* with norepinephrine administration: the more distal small diameter arterioles exhibiting a more pronounced response [123]. Autoregulatory escape is generally believed to be due to intrinsic metabolic mechanisms overriding the neurogenic constriction [62]. This contention is supported by the observations that (1) autoregulatory escape is more pronounced in the more metabolically active mucosa than the muscularis [124], (2) gastric oxygen consumption is maintained during autoregulatory escape in the stomach [125], and (3) blockade of adenosine bioavailability attenuates the autoregulatory escape [126]. Sympathetic stimulation also induces venular constriction in the intestine and mesenteric venules [122,123]. However, unlike the response in resistance vessels, autoregulatory escape does not occur in venules

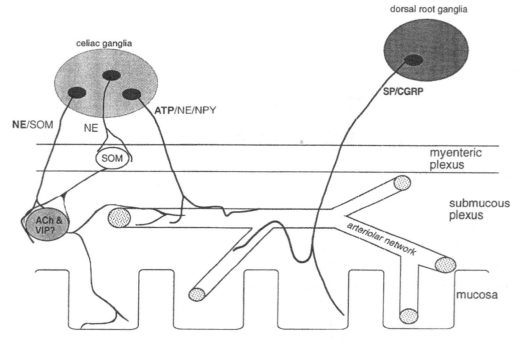

FIGURE 4.1: Schematic of the extrinsic and intrinsic innervations of submucosal arterioles in the guinea pig small intestine. NE, norepinephrine; SOM, somatostatin; Ach, acetylcholine; VIP, vasoactive intestinal polypeptide; NPY, neuropeptide Y; SP, substance P; CGRP, calcitonin gene-related peptide. Used with permission from *Am. J. Physiol.* 1996; 271: pp. G223–G230.

[127]. From a systemic homeostasis point of view, this selective autoregulatory escape of resistance vessels within the gastrointestinal microcirculation is important. For example, during blood loss, the sympathetic-induced sustained venoconstriction allows for expulsion of a significant volume of blood into the systemic circulation, while resistance vessel escape allows for sufficient oxygen delivery to maintain organ function.

4.1.2 Sensory C Fibers

Sensory C fibers, whose cell bodies are located in the dorsal root ganglia (Figure 4.1), also have efferent limbs impinging on submucosal arterioles. When these nerves are activated by ligands (capsaicin) of the vanilloid receptor (TRPV1), they induce dilation of the submucosal arterioles via the co-release of SP and CGRP [118,128]. Species and/or regional differences seem to exist along the gastrointestinal tract, i.e., CGRP is the predominant peptide in sensory neurons of the rat stomach

[129], while in the human colon, SP is the predominant neurotransmitter [130]. Further, although SP and CGRP have been the most studied neurotransmitters of these neurons, there is potential for a wider array of neurotransmitters [131]. CGRP can act directly on vascular smooth muscle to induce relaxation or can act via an interaction with the endothelium to induce an NO-mediated vasodilation [132]. Of interest is the fact that these capsaicin-sensitive C fibers are nociceptors responding to noxious chemical, mechanical, or thermal stimuli [133]. Thus, in addition to activating a local reflex vasodilation, they also pass on nociceptive information to the CNS.

In addition to the classical view that the afferent limbs of sensory C fibers are located within the mucosal interstitium, there is evidence that mesenteric resistance vessels may also be richly innervated with these C fibers [131,134]. This arteriolar neural reflex arc may modulate both the myogenic and metabolic intrinsic vasoregulatory mechanisms within the gastrointestinal tract. For example, the myogenic vasoconstriction in isolated mesenteric arteries is substantially blunted by ablation (capsaicin) of sensory C fibers, an effect also noted in arteries in which the endothelium was removed [134]. Based on pharmacologic and genetic blockade approaches, it was proposed that elevation of luminal pressure results in the generation of 20-HETE by smooth muscle, which subsequently activates TRPV1 (vanilloid receptor) on C fibers in the arteriole. The release of SP from the nerve terminals of these fibers interacts with NK1 receptors on smooth muscle cells causing contraction. This challenge to the current dogma that the myogenic response is an intrinsic property of vascular smooth muscle [52] cannot be readily dismissed, since a role for sensory C fibers has also been demonstrated in the myogenic response of isolated cerebral arteries [135]. C fibers have also been implicated in vasodilator responses in the gastrointestinal tract generally attributed to intrinsic metabolic regulatory mechanisms. Using capsaicin as a probe, these sensory C fibers have been implicated in reactive hyperemia in the intestine [136] and skin [137] as well as autoregulatory escape in the intestine [138] and the stomach [139]. These latter studies did not identify the neurotransmitters involved or characterize the reflex arc. Clearly, further studies are required to determine the circumstances under which these sensory fibers elicit vasodilation or vasoconstriction and why nociceptor nerves are involved in intrinsic vasoregulatory phenomenon.

4.1.3 Enteric Nerves

Gastrointestinal vasodilator neurons, whose cell bodies are located in the enteric ganglia, have been identified (Figure 4.1). The neurogenic vasodilation induced by these neurons is not affected by extrinsic denervation of the sympathetic supply or removal of the myenteric plexus [118,140]. These neurons contain acetylcholine (Ach) and vasoactive intestinal polypeptide (VIP). In the small intestine, the intrinsic vasodilation has been attributed to acetylcholine release from nerve terminals and subsequent NO generation from endothelium, while in the colon, VIP is the predominant neurotransmitter. These fibers can be activated by mechanical or chemical stimuli to induce vaso-dilation [118].

4.2   CIRCULATING VASOACTIVE SUBSTANCES

The major endogenous circulating substances that impact on gastrointestinal microvascular tone are catecholamines, vasopressin, and angiotensin II. Basal intestinal microvascular tone is attributed to the endogenous circulating levels of these substances. There may be some species variability in their relative contributions to vascular tone, i.e., norepinephrine and vasopressin being more important in the rat [141], while vasopressin and angiotensin II are predominant in the cat [142]. Further, there appears to be a redundancy in the vasopressin and angiotensin II systems, i.e., when one is inhibited, the other compensates to maintain tone [142].

Exogenous administration of catecholamines induces either vasoconstriction or vasodilation depending on whether the predominant effect is on α- or β-adrenergic receptors. Norepineph-rine, a predominantly α-adrenergic agonist, causes vasoconstriction and a decrease in intestinal blood flow [34,143,144]. As described above, during prolonged infusion of norepinephrine, the intestinal resistance vessels exhibit "autoregulatory escape." Epinephrine at low doses induces va-sodilation via β-adrenergic receptors and at high doses, vasoconstriction via α-adrenergic receptors [143,145,146].

In general, vasodilators increase $K_{f,c}$, while vasoconstrictors decrease $K_{f,c}$ [28,42]. For example, isoproterenol increases intestinal blood flow and $K_{f,c}$, the increase in blood flow being directly cor-related to the increase in $K_{f,c}$ [147]. Since isoproterenol does not affect capillary permeability, the increase in $K_{f,c}$ must be due to an increase in capillary exchange capacity (surface area). This ob-servation, coupled to the responses of other vasodilators and vasoconstrictors, indicate that vascular elements controlling capillary surface area (precapillary sphincters) possess specific receptors for a wide variety of vasoactive agents [42]. Some vasodilators can increase microvascular permeability (e.g., histamine, bradykinin), while some vasoconstrictors (e.g., angiotensin II) can decrease per-meability. Thus, unless estimates of capillary permeability are available, caution should be used in attributing changes in $K_{f,c}$ solely to corresponding changes in capillary surface area.

The effects of vasoactive agents on gastrointestinal oxygen uptake are variable and dependent on either their (1) direct effects on tissue metabolism or (2) indirect effects on oxygen delivery and/or capillary density [69]. Figure 4.2 depicts the relationship between gastrointestinal oxygen uptake and blood flow (or O_2 delivery) and circumstances that alter this relationship. Based on this theo-retical template, a vasodilator that increases oxidative metabolism will take pathway A. For example, dinitrophenol, an uncoupler of mitochondrial oxidative phosphorylation, increases both blood flow and oxygen consumption [66]. A vasodilator that does not affect oxidative metabolism (e.g., isopro-terenol) will take pathway B. Vasoconstrictors tend to reduce gastrointestinal oxygen uptake as de-

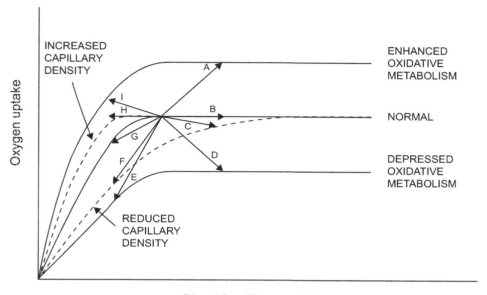

FIGURE 4.2: Relationship between oxygen uptake and blood flow (oxygen delivery). The curves depicted represent a composite of those shown in Figure 3.4. Alterations in tissue oxidative metabolism shift the curve vertically, while alterations in perfused capillary density shift the curve horizontally. The dot represents blood flow under control conditions, and the lettered arrows represent the potential effects of vasoactive agents on oxygen uptake. Pathway A is taken by a vasodilator that increases oxidative metabolism; Pathway B is taken by a vasodilator that does not affect oxidative metabolism or perfused capillary density; Pathway C is taken by a vasodilator that decreases capillary density; Pathway D is taken by a vasodilator that decreases metabolism; Pathway E is taken by a vasoconstrictor that decreases metabolism; Pathway F is taken by a vasoconstrictor that decreases capillary density; Pathway G is taken by a vasoconstrictor that does not affect metabolism or capillary density; Pathway H is taken by a vasoconstrictor that increases capillary density; Pathway I is taken by a vasoconstrictor that increases metabolism. Used with permission from *Am. J. Physiol.* 1982; 243: pp. G1–G9.

picted by pathway G, unless they also decrease capillary density, in which case they follow pathway F. For example, vasopressin and norepinephrine decrease blood flow, capillary density, and oxygen uptake [34]. By contrast, epinephrine can decrease blood flow and increase capillary density, yet not affect oxygen consumption (pathway H) [69]. Interestingly, after blockade of β-receptors, epinephrine decreased blood flow and oxygen uptake presumably by decreasing capillary density (pathway F) [145].

CHAPTER 5

Postprandial Hyperemia

5.1   GENERAL CHARACTERISTICS

It is well established that gastrointestinal blood flow increases after meals, a phenomenon referred to as postprandial or functional hyperemia [7,22,42,148,149]. The prevalence of this phenomenon is underscored by the demonstration of a postprandial hyperemia in man [150,151], monkeys [152], dogs [9,153–155], cats [156], rats [157], snakes [158], and fish [159]. In general, ingestion of food results in a dual hemodynamic response within the gastrointestinal tract: an initial transient response during anticipation and ingestion of food and a subsequent prolonged response during digestion and absorption.

The anticipatory/ingestion phase is characterized by increases in heart rate, cardiac output, and aortic pressure with only minor changes in gastrointestinal vascular resistance [153,154]. These transient hemodynamic alterations can be (1) blunted by adrenergic blocking agents and (2) mimicked by allowing the animals to see and smell the food, but not allowed to ingest it [153]. Thus, this transient phase represents a sympathetic-driven cephalic phase.

The digestive/absorptive phase is characterized by a gastrointestinal hyperemia. In conscious animals, blood flow in the left gastric, celiac, and superior mesenteric arteries increases within minutes after ingestion of a meal [153,154,160,161]. Left gastric and celiac artery blood flow increases earlier and is transient (10–15 min), while superior mesenteric artery blood flow increases later and is more prolonged (up to several hours). The postprandial hyperemia is detected earlier in the jejunum (within 30 min) than the ileum (by 90 min) [9]. Collectively, these observations indicate that the postprandial hyperemia progresses along the gastrointestinal tract in association with the aboral movement of ingested food. The magnitude (25–200%) and the duration (3–7 h) of the hyperemia appear to depend on the composition of the meal. In general, lipid- and protein-rich meals are more potent than carbohydrate-rich meals in eliciting a hyperemia [152,156]. A cholinergic neural pathway has been proposed to be involved in the gastrointestinal postprandial hyperemia [152,153], but this neurogenic-mediated contribution to the hyperemia may be indirect, rather than direct (see below). In general, the characteristics of the gastrointestinal postprandial hyperemia in animals mimics those observed in humans [162,163].

5.2   LOCALIZATION OF THE POSTPRANDIAL HYPEREMIA

The bulk of experimental evidence indicates that the postprandial hyperemia is localized to that portion of the canine gastrointestinal tract exposed to food or hydrolytic products of food digestion [42,149,164]. Intragastric placement of undigested food increases celiac artery blood flow within minutes, followed by an increase in SMA blood flow [165]. Intrajejunal placement of digested food does not increase celiac artery blood flow, but increases SMA blood flow. These findings are in accord with observations in conscious animals. Further, intrajejunal placement of digested food or hydrolytic products of food digestion increases local blood flow without effecting blood flow to an adjacent segment [165,166]. In general, the nutrient-induced increase in intestinal blood flow appears to be confined to the mucosal layer of the gut wall [9,165,167,168]. However, there is experimental evidence to indicate that intraluminal food (or nutrients) increases blood flow to segments of the small bowel not exposed to chyme [155,156] and that the hyperemic response involves both the mucosal and muscularis layers of the bowel wall [84,157]. The reasons for the discrepancies are not clear, but a role for reflex motor activity in response to luminal nutrients is worth appraisal.

5.3   CONSTITUENTS OF CHYME RESPONSIBLE FOR THE POSTPRANDIAL HYPEREMIA

The constituents of chyme responsible for the postprandial hyperemia in the small intestine and colon are summarized in Figure 5.1 [42]. The constituents of chyme that elicit a local hyperemic response differ in different regions of the intestine. In the upper small intestine (jejunum), intraluminal placement of digested food, but not undigested food, increases local blood flow (Figure 5.1), with a high fat test meal producing the greatest hyperemic response followed by the high protein and high carbohydrate diets [166,169]. Collectively, these findings indicate that the hydrolytic products of food digestion are responsible for the postprandial hyperemia in the upper small intestine.

Intrajejunal placement of amino acids or peptides (at postprandial concentrations) did not affect local blood flow [166]. By contrast, intrajejunal glucose solutions produced a hyperemic response [166,170]. Emulsions of oleic acid (long-chain fatty acid) did not affect blood flow, unless oleic acid was solubilized with bile or bile salts, in which case, it produced a significant hyperemic response [166,171]. The hyperemic response to micellar oleic acid was dose-dependent; 40 mM oleic acid was twice as potent as 20 mM oleic acid. Bile-solubilized caproic acid (short-chain fatty acid) did not induce a hyperemia [172]. Bile (or bile salts) or pancreatic enzymes were ineffective in eliciting a hyperemic response. Collectively, these findings indicate that solubilized hydrolytic products of long-chain lipids and carbohydrates are the primary mediators of the intestinal postprandial hyperemia. However, based on the effects of mixed diets, a synergistic effect of all the hydrolytic products of food digestion cannot be discounted.

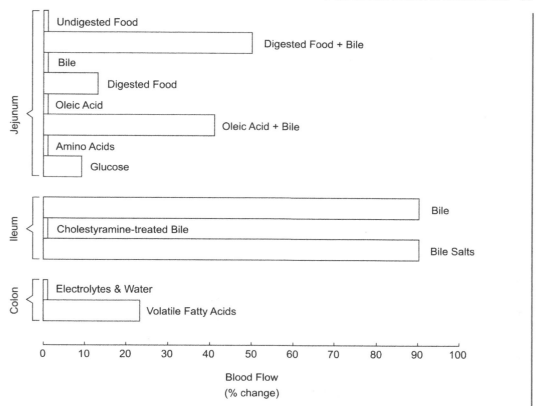

FIGURE 5.1: Effects of intraluminal placement of various constituents of chyme on intestinal blood flow. Used with permission from *Handbook of Physiology, The Gastrointestinal System I*, Chapter 39, 1989, pp. 1405–1474.

In addition to playing a permissive role in the oleic acid-induced hyperemia in the jejunum, bile appears to play an important direct role in the local postprandial hyperemia of the ileum (Figure 5.1). In the ileum, bile induces an increase in blood flow, the effect being mediated by bile salts. This contention is supported by the observations that (1) bile salts, rather than other constituents of bile, can induce a hyperemia similar to that seen with bile and (2) cholestyramine, a bile salt-sequestering agent, prevents the bile-induced ileal hyperemia [37].

A significant amount of carbohydrates (e.g., dietary fiber) can escape absorption in the small intestine and reach the colon where bacterial fermentation converts them to volatile fatty acids. The major volatile fatty acids present in the colon are acetic, propionic, and butyric acids. Intracolonic placement of acetic acid increased local colonic blood flow; butyric or propionic acids were without effect [173].

5.4 MECHANISMS INVOLVED IN THE POSTPRANDIAL HYPEREMIA

Secretory, absorptive, and motor activity are all important functions of the gastrointestinal tract required for efficient assimilation of ingested nutrients. Thus, the mechanisms involved in the postprandial gastrointestinal hyperemia may be quite complex and are not as yet clearly defined. However, several potential mechanisms have been implicated in the hyperemia, including extrinsic and enteric nerves, circulating hormones, and tissue metabolic activity [22,42,149,164].

5.4.1 Extrinsic Nerves

Aside from the initial systemic hemodynamic response during the anticipation/ingestion of food, adrenergic blockade does not influence the postprandial gastrointestinal hyperemia. However, cholinergic nerves have been proposed to be involved in the gastrointestinal postprandial hyperemia [22,42]. The specific neural pathways may be different for the hyperemic response in the stomach and intestines. Vagotomy, but not atropine, blocks the gastric (celiac artery) hyperemia [160,161], while atropine, but not vagotomy, blocks the intestinal (superior mesenteric artery) hyperemia [22,152,153,160]. The role of cholinergic nerves in the gastrointestinal hyperemia, however, may be indirect. Effective assimilation of ingested food is a multistep process. In the stomach, food is partially digested by gastric acid and lipases and then emptied into the small intestine. In the small intestine, progressive digestion and solubilization of the hydrolytic products is facilitated by pancreatic and biliary secretions. All of these processes can be severely blunted by cholinergic blockade. Thus, the cholinergic modulation of the gastrointestinal hyperemia may simply be a result of the inhibition of appropriate processing and delivery of nutrients to absorptive sites. In support of this possibility are the observations that (1) introduction of solubilized products of food digestion into the upper small intestine induces a local hyperemia, while undigested food does not [166], extrinsic denervation of the intestine does not prevent the intestinal hyperemia induced by luminal nutrients [174] and (2) atropine does not block the hyperemic response to predigested food [174,175].

5.4.2 Enteric Neural Reflexes

There also appears to be some support for the role of local enteric nerves in the postprandial hyperemia. The postprandial gastric hyperemia is prevented by anesthetizing (oxethazaine) the gastric mucosa [161]. Mechanical stimulation of the small intestinal mucosa elicits a local vasodilation that is mediated by a noncholinergic, nonadrenergic (NANC) enteric reflex, which can be blocked by tetrodotoxin [176,177]. The mechanical stimulation of the mucosa with a plastic object has been taken to simulate the movement of chyme along the intestine [149]. However, the results of studies with digested food, or hydrolytic products thereof, cast doubt on the potential role of enteric nerves in the postprandial intestinal hyperemia. As mentioned above, intrajejunal placement of digested food, but not undigested food, induces a local hyperemic response [166]. Thus, unlike mechanical stimulation of the mucosa with coarse foreign material, a simple mechanical interaction of foodstuff with the mucosa does not induce a hyperemia. Introduction of digested food, glucose, or micellar oleic acid into isolated jejunal segments produces a local hyperemia, which is not affected by tetrodotoxin or hexamethonium [174]. The ability of local anesthesia of the mucosa to inhibit the local hyperemia to hydrolytic products of food digestion may also be indirect. For example, the hyperemic response to hydrolytic products of food digestion could be blocked by dibucaine, an effect attributed to the inhibition of transport processes of the mucosa rather than decreased activity of local nerves [80,174].

5.4.3 Circulating Hormones

Various gastrointestinal hormones and neuropeptides are released into the circulation postprandially, including gastrin, secretin, cholecystokinin (CCK), vasoactive intestinal polypeptide (VIP), neurotensin, and substance P. Since all of these peptides are vasodilators in the gastrointestinal circulation, it has been proposed that one or more of them may contribute to the postprandial hyperemia [22,149,156]. However, in general, their role in the postprandial intestinal hyperemia has been dismissed based on the fact that their vasodilator effects are only apparent when they are administered at doses 10–100 times greater than those measured in the circulation postprandially [149,178–181]. However, circulating levels of these hormones may only represent a fraction of the levels in the interstitium where they are generated adjacent to intestinal resistance vessels. To illustrate this point is the case of VIP [181]. To elicit a jejunal hyperemia, VIP had to be administered at 10 times the concentration measured in the venous effluent from jejunal segments containing micellar oleic acid. Yet, antisera to VIP reduced the hyperemic response to micellar oleic acid. Another reason for eliminating a role for a given hormone released after meals is its effects on intramural blood flow distribution [149]. For example, intravascular administration of neurotensin to achieve postprandial concentrations, increased muscularis, but not mucosal, blood flow in the ileum [182]. This is analogous to excluding adenosine as potential mediator of metabolic regulation based on arterial infusion to a multicompartmental organ (see 3.4.2). Thus, although it is generally accepted that circulating gastrointestinal peptides play a minor role (if any) in the postprandial hyperemia, their potential paracrine contribution has not been systematically addressed.

5.4.4 Tissue Metabolic Activity

An increase in splanchnic oxygen consumption (O_2 demand) and blood flow (O_2 delivery) after a meal has been demonstrated in man [151]. The increase in blood flow is presumed to be a result of the increased O_2 requirements for efficient digestion and absorption of ingested food. Secretory and absorptive activity is generally confined to the mucosa, while propulsive motor activity is the responsibility of the muscularis. In general, gastric mucosal blood flow increases with enhanced acid secretion [78] and intestinal mucosal blood flow increases during absorption of nutrients, [166] while muscle blood increases during enhanced rhythmic contractions of the gut [11]. Thus, the postprandial (functional) gastrointestinal hyperemia is a rather complex phenomenon involving alterations in the relative distribution of blood flow between the mucosal and muscularis compartments at any given time. Thus, investigative approaches have tended to isolate these activities (solute transport vs motility) when addressing the role of metabolic factors in the postprandial hyperemia.

Solute Transport. Gastric oxygen consumption (O_2 demand) increases during stimulation of acid secretion by secretagogues (e.g., gastrin, histamine) [65,78,183,184]. Oxygen uptake of the small and large intestine increases during absorption of nutrients or transportable solutes [56,76,80,173]. Indeed, a direct correlation between the rate of solute transport and local oxygen consumption has been demonstrated in isolated preparations of the stomach [184,185], small intestine [39,170], and colon [56].

In most cases, there is an increase in regional blood flow associated with enhanced functional activity; however, there are some exceptions. While histamine-induced gastric acid secretion is associated with an increase in gastric blood flow, pentagastrin-induced gastric acid secretion may not be associated with a local hyperemic response [186]. Similarly, glucose absorption by the intestine can increase oxygen uptake, yet have variable effects on blood flow [39]. According to the metabolic theory, an increase in tissue oxygen demand can be met by changes in either blood flow (resistance vessel regulation) or oxygen extraction (exchange vessel regulation). One determinant of whether resistance vessel or exchange vessel regulation predominates is the level of existing oxygen extraction or $(A - V)O_2$ (Figure 5.2) [44]. If the prevailing $(A - V)O_2$ is low, then the increase in postprandial oxygen demand is met primarily by increases in O_2 extraction. If the prevailing $(A - V)O_2$ is high, then the postprandial oxygen demand is met by increases in intestinal blood flow. Another confounding factor is the vasoactive properties of the agent used to enhance functional (metabolic) activity. For example, histamine is a potent secretagogue and vasodilator, and increasing doses of histamine increase both gastric oxygen uptake and blood flow. Thus, according to the relationship between oxygen uptake and blood flow (or O_2 delivery), histamine should follow pathway A of the relationship depicted in Figure 4.2.

According to the metabolic theory, the link between oxygen demand and intrinsic regulation of resistance and exchange vessels is provided by local changes in tissue pO_2 and/or adenosine. Application of glucose to the mucosal surface reduces villus pO_2 and increases blood flow to the villus via the submucosal arteries [84]. Further, the glucose-induced hyperemia is diminished when villus pO_2 is stabilized by exogenous delivery of O_2 via the luminal aspect of the mucosa [83]. However,

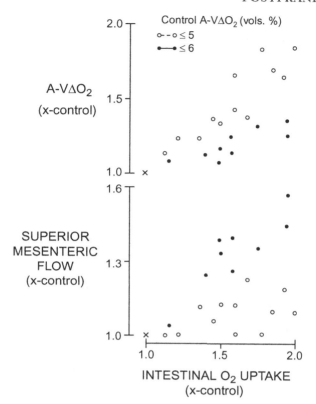

FIGURE 5.2: Relationships between changes in intestinal arteriovenous oxygen difference (upper panel) and blood flow (lower panel) in response to increases in oxygen demand imposed by instillation of digested food in the lumen. Solid circles represent responses when the initial arteriovenous oxygen difference $((A\text{-}V)O_2)$ was 6 vol % or greater. Open circles represent responses when $((A\text{-}V)O_2)$ was 5 vol % or less. The control point for each variable is represented by x. Note that the postprandial hyperemia was greater in animals with a higher initial $((A\text{-}V)O_2)$. Used with permission from *Am. J. Physiol.* 1980; 238: pp. H836–H843.

submucosal arteries also dilate during luminal application of glucose, even though perivascular pO_2 in this region is unchanged. In the case of adenosine, it is released from the small intestine (into the draining vein) when digested food is placed in the lumen [88]. However, blockade of adenosine bioactivity produced equivocal results with respect to attenuating the hyperemic response to luminal digested food [89,90].

NO has also been implicated in the postprandial gastrointestinal hyperemia. Blockade of eNOS blunts both the gastric hyperemia associated with pentagastrin-induced acid secretion [97,106] and the intestinal hyperemia associated with glucose absorption [187]. In the small intestine, glucose absorption is associated with increases in NaCl hyperosmolarity at the level of the sub-mucosal arterioles and larger feeding arteries [188]. The hyperosmolarity-induced vasodilation in the submucosal vessels is partially inhibited by blockade of eNOS [104]. Based on pharmacologic blockade approaches, it has been proposed that an increase in intracellular Ca^{++} levels via the Na^+/Ca^{++} exchanger is crucial for the hyperosmolarity-induced increase in endothelial cell production of NO [105]. Collectively, the information available on the role of NO in the absorptive hyperemia provides a potential feedback link by which microvascular adjustments are regulated during inter-mittent periods of absorptive activity. Specifically, the increased NaCl hyperosmolarity would lead to NO production and resistance vessel dilation. The vasodilation would then wash out the hyper-osmolarity and remove the stimulus for NO production, thereby closing the feedback loop.

Motor Activity. Although in isolated visceral smooth muscle (*Taeniae coli*) enhanced contractile activ-ity is associated with increases in oxygen consumption [189], in isolated perfused preparations of the small intestine, increases in motility have unpredictable effects on blood flow and oxygen consump-tion [190–192]. This may be a result of the capricious nature of motor activity in the small intestine, being of the rhythmic or tonic type or a combination of both at any given time. Rhythmic contrac-tions can passively affect blood flow through intermittent compression of blood vessels, decreasing blood flow during the contractile phase and increasing blood flow during the relaxation phase [193]. Tonic contractions tend to decrease small intestinal blood flow due to vessel compression. When the influence of tonic contractions is minimized, a positive correlation is obtained between the level of intestinal motility and oxygen consumption [38]. When a vasoconstrictor (met-enkephalin) was used to stimulate motility, the increase in oxygen uptake was met entirely by an increase in $(A - V)O_2$. Alternatively, when a vasodilator (acetylcholine) was used, the increase in oxygen demand was met by increases in blood flow. Thus, according to the relationship between oxygen uptake and blood flow (or O_2 delivery) depicted in Figure 4.2, met-enkephalin follows pathway I, while acetylcholine follows pathway A.

Little is known regarding the mediators which link the oxygen requirements of enhanced motor activity to the regulation of resistance or exchange vessels. However, based on studies in other vascular beds supplying muscular tissue, such as heart and skeletal muscle [194,195], tissue pO_2, adenosine, and NO may be likely candidates. Studies are warranted to directly address this issue.

. . . .

CHAPTER 6

Transcapillary Solute Exchange

6.1 ULTRASTRUCTURAL PATHWAYS

The exchange vessels (capillaries) of the gastrointestinal mucosa are fenestrated and relatively permeable to small molecules (e.g., glucose), yet restrict the transcapillary movement of large molecules (e.g., albumin). This allows for the absorption of hydrolytic products of food digestion without compromising the oncotic pressure gradient governing transcapillary fluid movement and edema formation. Ultrastructural studies (e.g., electron microscopy) have identified potential pathways for solute and water exchange across these fenestrated capillaries (Figure 6.1) [28]. Similar pathways for solute and water movement exist in the continuous capillaries of the muscularis, with the obvious exception of fenestrae. A brief description of the enumerated pathways depicted in Figure 6.1 is as follows.

6.1.1 Endothelial Cell Membrane

Lipid-soluble substances (O_2 and CO_2) and very small nonpolar hydrophilic substances (urea) can traverse the fenestrated capillaries by diffusing through the cell membrane proper. The permeability of the endothelial cell has been described in terms of small perforations (4–10 Å radius) in the luminal and abluminal lipid bilayers and assuming negligible resistance to diffusion offered by the cytoplasm. This pathway accounts for less than 10% of the capillary hydraulic conductivity

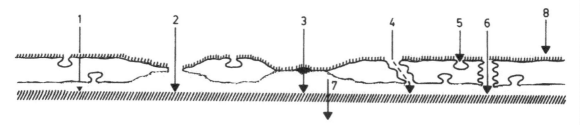

FIGURE 6.1: Schematic of solute and fluid transport pathways in the capillaries of the gastrointestinal tract. Transport pathways: 1, cell membrane; 2, open fenestrae; 3, diaphragmed fenestrae; 4, intercellular junction; 5, pinocytotic vesicles; 6, transendothelial channels; 7, basement membrane; 8, glycocalyx. Modified from and used with permission from *Gastroenterology* 1983; 84: pp. 846–868.

(transcapillary volume flow rate per unit pressure gradient). Thus, the bulk of the transcapillary fluid and solute movement occurs through extracellular pathways [28,42,196].

6.1.2   Fenestrae

Fenestrae are circular openings (200 to 300 Å in radius) in the capillary endothelium. There is an asymmetry in the relative number of fenestrae both along the length of the capillaries as well as with respect to transporting epithelia. The frequency of the fenestrae increases from arterial to venous ends of the capillary, and they are preferentially oriented toward the base of the transporting epithelia. Further, in the small intestine, the capillaries of the villus tips and the crypt region contain more fenestrae per cross-section than the capillaries located at the base of the villus. Tracer molecules ranging in size from 25 to 120 Å can pass through the fenestrae. Approximately half of the fenestrae are closed by a diaphragm, which is believed to offer some restriction to the movement of solutes [12,28,197–202].

6.1.3   Pinocytotic Vesicles

The capillary endothelium also contains plasmalemmal (pinocytotic) vesicles with an internal radius of approximately 250 Å. As is the case with fenestrae, their frequency increases from the arteriolar to venular end. The vesicles are believed to be mobile structures opening on one side of the endothelium (caveolae), acquiring solutes and fluid, moving across the endothelial cytoplasm and discharging their contents on the opposite side. Solutes 25 to 150 Å are readily transported by these vesicles; however, the overall transport of macromolecules by vesicular transport is much slower than their movement through fenestrae. One or more vesicles (caveolae) can fuse and form an open transendothelial channel; however, they have strictures at the points of fusion that reduces their internal radius from 250 Å to between 50 and 200 Å. The effective radius is further reduced in some channels by the presence of diaphragms at the points of fusion [197,201–204].

6.1.4   Interendothelial Cell Junctions

The intercellular junctions can be either open or closed; the frequency of open intercellular junctions increases from the arteriolar to the venular end of the capillaries. The open junctions behave as if they are channels 20 to 60 Å in width, while the closed channels are functionally impermeable to solutes of 20 Å diameter [42,199,202].

6.1.5   Glycocalyx

An extra-endothelial barrier to solute movement may be the glycocalyx on the luminal endothelial cell surface [114]. In the fenestrated capillaries of the gastrointestinal mucosa, the glycocalyx not only covers the endothelial cell proper, but also appears to line the luminal portion of the interendo-thelial junction and covers the surface of fenestral diaphragms [205–207]. Enzymatic removal of the glycocalyx (e.g., pronase) can decrease the hydraulic conductivity of fenestrated capillaries [208].

6.1.6 Basement Membrane

The basement membrane surrounding the endothelium also possesses restrictive properties. Tracers of molecular radius ranging from 25 to 150 Å can readily cross the endothelial lining. Tracers 25–55 Å in radius do not appear to be impeded in their further movement across the basement membrane. Molecules of between 62 and 150 Å are temporarily delayed at the level of the basement membrane, at times accumulating in clusters under permeable fenestrae [28,197,199,209].

6.2 PHYSIOLOGICAL (FUNCTIONAL) PATHWAYS

Fluid exchange can occur through any of the multiple small ultrastructural "channels," but exchange of solutes and proteins of different molecular weights will be strictly dependent on channels whose diameters exceed those of the solutes/proteins. In addition to the permeability (porosity) of the capillaries, transcapillary movement of solutes can be influenced by the capillary surface area. Thus, a major physiological approach to this issue has been to assess the transcapillary movement of solutes of different molecular weights (radii), while minimizing the effects of capillary surface area. To this end, a variety of techniques have been used to assess gastrointestinal microvascular permeability to solutes, each with its own set of advantages and limitations [204,210]. Of the approaches used to assess the permeability of the gastrointestinal tract microcirculation, the indicator dilution technique to study small (<37 Å) solute permeability and the steady-state analysis of lymph protein flux to assess macromolecular (>37 Å) permeability have yielded the most useful information [28,42,197]. Mathematical modeling approaches using experimentally and theoretically derived information have yielded estimates of "equivalent pore radii" to describe the functional permeability characteristics of the gastrointestinal capillaries [196,204,210].

6.2.1 Small Solutes

The multiple indicator dilution technique uses radioactive diffusible tracer molecules to assess their extraction by capillaries during a single pass through the capillary bed [211]. Pairs of diffusible tracers and a nondiffusible (vascular tracer) are simultaneously injected into the arterial supply of isolated organs, and the relative concentrations of the tracers in the venous effluent are assessed. The relative concentration of an isotope in venous blood is obtained by expressing the concentration relative to the original concentration in the injectate. The extraction (E) of a diffusible tracer is calculated by

$$E = [C_V(t) - C_D(t)] / C_V(t)$$

where $C_V(t)$ equals the relative concentration of the vascular tracer in the venous sample and is equivalent to the relative concentration of the diffusible tracer in arterial blood at time t, and $C_D(t)$ equals the relative concentration of the diffusible tracer in venous blood at time t.

When plasma flow (blood volume flow corrected for Hct) to the isolated preparations of the stomach and small intestine are measured, the permeability–surface area product (PS) for the diffusible tracers can be calculated by

$$PS = Q_P \ln(1 - E)$$

where Q_p is plasma flow through the preparation.

The use of multiple tracers of different sizes has yielded information relevant to the permselectivity of the gastric [212] and small intestinal [213] capillaries. As mentioned above, the mucosal and muscularis microcirculations are arranged in parallel, and the capillaries of the mucosa have fenestrae, while those of the muscularis do not. Since the mucosal circulation receives approximately 80% of the intramural blood flow, the permeability of the mucosal capillaries will dominate the outcome of the experiments. In these studies, isoproterenol was used to preferentially increase mucosal blood flow, increasing the probability that the experimental outcome reflected primarily the permeability characteristics of the mucosal capillaries. Under these conditions, the ratios of the PS values for simultaneously injected inulin (15 Å) and β-lactoglobulin (28 Å) were greater than the ratios of their free diffusion coefficients. This indicates that the extraction of β-lactoglobulin was more restricted by the capillary wall than that of inulin. Since the diffusible tracers are injected simultaneously, their PS ratios are assumed to be the same as their permeability ratios, given that the surface area is the same for both tracers and cancels out. Thus, based on the relationships among the permeabilities, diffusion coefficients, and radii of these two molecules, [214] an equivalent small pore radius of 53 and 59 Å is predicted for the mucosal capillaries of the stomach [212] and small intestine, [213] respectively.

6.2.2 Macromolecules

Steady-state analysis of lymphatic protein flux has been used to provide information on the permeability of gastrointestinal capillaries to macromolecules (e.g., endogenous plasma proteins) [210,212,215]. A major assumption of this approach is that the concentration of macromolecules in the lymph draining the gastrointestinal tract is identical to that of interstitial fluid [204]. The validity of this assumption is based on both ultrastructural studies of initial lymphatics indicating that the endothelium is discontinuous and the basement membrane fragmented [198] and physiologic studies indicating that the interstitial and lymph protein concentration is similar [216]. Further, complicating the lymphatic protein flux approach is the anatomical regions drained by the lymphatics. The draining lymphatics in the mesentery contain contributions from both the mucosal and muscularis regions. However, the protein concentration of the rat villus lacteal and the collecting lymphatics are similar [217]. Collectively, these observations support the contention that protein concentration profile of collecting lymphatics provides a reasonable estimate of that in the mucosal interstitial fluid.

Under resting conditions (normal microvascular pressures), transcapillary exchange of macromolecules occurs by both convection and diffusion. Thus, an assessment of the lymph-to-plasma protein concentration ratio (C_L/C_P) does not allow for an accurate approximation of the sieving characteristics of capillaries. In order for a lymph protein flux analysis to yield useful information regarding the permeability characteristics of the capillaries, the convective flux must be maximized and the diffusive flux minimized. An experimental approach, which increases the convective movement of macromolecules across the capillaries to such an extent that their diffusive exchange becomes negligible, is acute venous hypertension. When capillary filtration (or lymph flow) is increased by graded venous hypertension, C_L/C_P for total plasma proteins progressively decreases (Figure 6.2). This portion of the relationship is termed "filtration rate-dependent." At high lymph flows, C_L/C_P reaches steady-state levels, i.e., becomes "filtration rate-independent." When C_L/C_P is filtration rate (or lymph flow)-independent, lymph protein flux analyses can provide information on the true sieving characteristics of the capillary wall. Specifically, the level at which C_L/C_P becomes filtration rate-independent can provide a reasonable estimate of the osmotic reflection coefficient, σ_d (1 – C_L/C_P in Figure 6.3) [210,215].

The relationship between total protein C_L/C_P and lymph flow rate depicted in Figure 6.2 was derived from studies in the cat and rat small intestine. Similar relationships have been obtained in the feline stomach and canine small intestine and colon. The σ_d values for total protein and various endogenous proteins of different sizes in the fenestrated capillaries of the gastrointestinal tract [30,212,215,218] are compared to corresponding σ_d values obtained for the continuous capillaries of the hindpaw [219] in Table 6.1. As shown in Table 6.1, σ_d increases as the radii of the endogenous proteins increase, indicating permselectivity of gastrointestinal capillaries. Based on the available data, species differences may exist, and caution should be exercised in making comparisons across species. Further, in the dog, σ_d values for total protein in the small and large intestinal capillaries are similar and not dramatically different from the continuous capillaries of the hindpaw.

The σ_d values for different-sized macromolecules may be used to quantitatively describe the permeability characteristics of the gastrointestinal microvasculature in terms of "equivalent pores" using a graphic analysis [204,210,220]. An example of such an analysis for the small intestine [215] is shown in Figure 6.4, where 1 – σ_d is plotted as a function of solute radius. The analysis involves

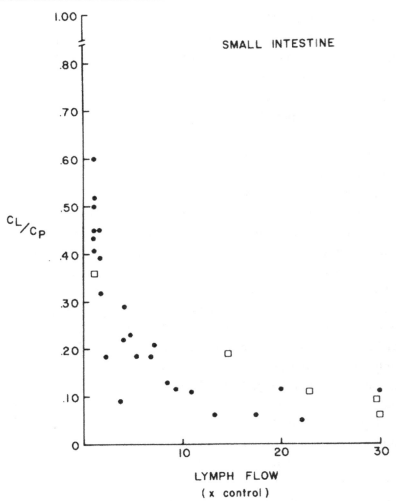

FIGURE 6.2: Relationship between lymph-to-plasma ratio for total protein concentration (L/P) and lymph flow. Intestinal lymph flow was increased by graded increases in venous pressure in cats (circles) and rats (squares). Used with permission from *Handbook of Physiology, The Gastrointestinal System I,* Chapter 39, 1989, pp. 1405–1474.

fitting the data with two sets of equivalent pores: small and large. First, a theoretical large-pore curve is fitted to the data points representing the large molecules. Subsequently, by a "curve-peeling" process, the resulting values for $1 - \sigma_d$ for small solutes are fitted with a smaller theoretical pore curve. The ordinate intercept predicts the percentage of the total hydraulic conductance occurring through each set of pores. The relative areas of the small and large-pore populations can be assessed using

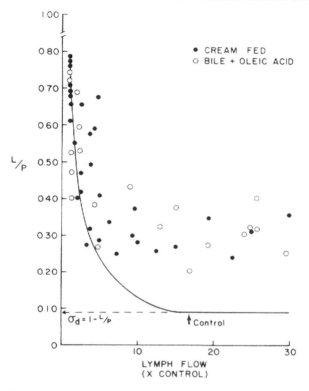

FIGURE 6.3: Relationship between the ratio of lymph to plasma total protein concentration (L/P) and lymph flow in the small intestine during graded acute venous hypertension. The solid line depicts the control situation (nontransporting small bowel). The experimental data obtained when lipids were present in the lumen indicate that σ_d (1 − L/P) is decreased to 0.7 from a control value of 0.9; i.e., intraluminal lipids increase capillary permeability to plasma proteins. Used with permission from *Am. J. Physiol.* 1982; 242: pp. G194–G201.

$$A_s/A_l = (F_s/F_l) \times [(r_l)^2 / (r_s)^2]$$

where A equals pore area, F equals the fraction of hydraulic flow through the pores, r equals pore radius, and the subscripts s and l refer to small and large pores, respectively. The relative frequency of each type of pore can be calculated by

$$N_s/N_l = (A_s/A_l) \times [(r_l)^2/(r_s)^2]$$

where N_s/N_l is the ratio of small-to-large pore number.

TABLE 6.1: Osmotic reflection coefficients[1] for different-sized proteins in the feline and canine gastrointestinal tract.

RADIUS (Å)	STOMACH (cat)	INTESTINE (cat)	INTESTINE (dog)	COLON (dog)	HINDPAW (dog)
Total	0.78	0.92	0.85	0.85	0.90
37	0.73	0.90		0.75	0.87
38	0.77	0.92	0.82		
39	0.78	0.94			
40			0.88	0.82	0.89
42	0.79	0.96	0.80		
44			0.84	0.87	0.89
48				0.88	
96	0.91	0.98			
100				0.95	0.96
≥120	0.91	0.99	0.99	0.98	0.97

[1]The osmotic reflection coefficient was calculated from C_L / C_P data obtained at high capillary filtration rates ($1 - C_L / C_P$). Data for the cat stomach are from Am. J. Physiol. 1981; 241: pp. G478–G486, for the cat intestine are from Am. J. Physiol. 1980; 238: pp. H457–H464, for the dog intestine are from Hypertension 1984; 6: pp. 13–19, for the dog colon are from Am. J. Physiol. 1980; 239: pp. G300–G305, and dog hindpaw are from Microvasc. Res. 1983; 26: pp. 250–253.

This type of mathematical analysis has also been applied to data obtained for the stomach [212] and colon [218] and cumulatively presented in Table 6.2. Several general features of the permselectivity of the gastrointestinal capillaries can gleaned from this information. The effective radius of the small and large pores do not vary dramatically between various regions of the gastrointestinal tract (15–30%), and the radius of the small pores derived using lymph protein flux data is similar to that predicted by the multiple indicator dilution technique, i.e., 53–59 Å. Further, the relative density (number and area occupied) by the small-pore component is much greater than the large-pore component, the disparity being the greatest in the capillaries of the small intestine. Also

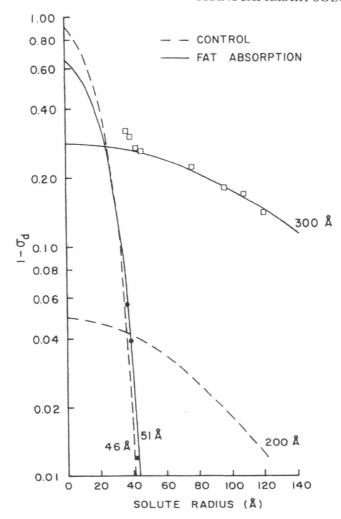

FIGURE 6.4: Application of pore-stripping analysis to σ_d values for plasma proteins of different sizes. The data are fitted using two sets of equivalent pores. Dashed lines depict analysis of data from nontransporting small intestine, while solid lines depict analysis of data obtained during fat absorption. First, a theoretical large pore curve is fitted to the data points representing the large molecules (open symbols). Subsequently, by a "curve-peeling" process, the resulting values for $1 - \sigma_d$ for small solutes are fitted with a smaller theoretical pore curve (closed symbols). Based on this analysis, fat absorption is associated with an increase in the size of the large pores (200 to 300 Å) while not affecting the small pores. Used with permission from *Am. J. Physiol.* 1982; 242: pp. G194–G201.

TABLE 6.2: Dimensions and relative frequencies of equivalent pores in the fenestrated capillaries of the gastrointestinal tract compared to continuous capillaries of the hind-paw.

ORGAN	SPECIES	SMALL-PORE RADIUS (Å)	LARGE-PORE RADIUS (Å)	A_s / A_l	N_s / N_l
Stomach	cat	47	250	92/1	2,600/1
Intestine	cat	46	200	340/1	6,400/1
Colon	dog	53	180	48/1	550/1
Hind-paw	*dog*	*47*	*195*	*114/1*	*2,064/1*

A_s / A_l = total area of small pores compared with total area of large pores; N_s / N_l = relative number of small pores compared with large pores. Modified and used with permission from *Pathophysiology of the Splanchnic Circulation, Volume I*, 1987; Chapter 1, pp. 1–56.

of interest is that the dimensions and relative frequencies of equivalent pores in the fenestrated capillaries of the gastrointestinal tract are similar to the continuous capillaries of the hind-paw.

In general, agents or conditions that increase intestinal capillary permeability do so by increasing the dimensions of the large pores [210]. One condition that appears to selectively influence the small-pore population is severe arterial hypoxemia (pO_2 of 35 mmHg) [221]. In this situation, the small pore size was increased (from 59 to 67 Å) as assessed by the multiple indicator dilution technique. However, lymph protein flux did not change, indicating that the dimensions of the large pores were not affected.

The capillaries of the gastrointestinal tract may also exhibit charge selectivity. Studies in the small intestine indicate endogenous lactate dehydrogenase (LDH; 42 Å radius) isoenzymes are selectively restricted by the capillaries based on their isoelectric points [222]. The σ_d for LDH isoenzymes decreased from 0.95 for the most positive isoenzyme (isoelectric point, 8.3) to 0.71 for the most negative isoenzyme (isoelectric point, 5.2). Similar results were noted using exogenously administered dextran molecules (35–36 Å radius), i.e., the steady-state C_L/C_P of the neutral dextran was twice that of the positive dextran. Other studies assessing binding of cationized ferritin to intestinal mucosal capillaries demonstrated that very little binding of this cationic molecule occurs, indicating charge repulsion by the endothelium [223]. Collectively, these observations indicate that the small intestinal microvessels behave as a positively charged barrier. This is in stark contrast to the renal glomerular microcirculation, which behaves as a negatively charged barrier [224]. It has been suggested that the cationic nature of the small intestinal capillaries promotes transcapillary movement of negatively charged proteins, such as albumin, which is ultimately returned to the circulation via lymphatics (and capillaries) [42]. This would favor assimilation of absorbed substances, which bind to albumin (e.g., fatty acids). By contrast, the anionic nature of the glomerular capillaries would prevent the urinary loss of negatively charged proteins [224].

6.3 FACTORS INFLUENCING VASCULAR PERMEABILITY

Several experimental interventions to mimic physiological and pathological situations have been shown to alter the permeability of small intestinal capillaries to endogenous plasma proteins (Table 6.3) [42,197].

Simulation of the postprandial state by intraluminal placement of micellar fatty acids or glucose and electrolytes is associated with an increase in transcapillary protein movement as reflected by an increase in lymphatic protein flux [225,226]. This increased protein flux may be the result of an increase in capillary surface area and/or an increase in capillary permeability to proteins. As shown in Table 6.3, the σ_d for total protein is not affected by glucose or electrolyte absorption, indicating that an increase in capillary surface area, rather than an increase in capillary permeability, is responsible for the increased lymphatic protein flux. This contention is supported by the observations that $K_{f,c}$ [227] and the PS for rubidium [71] is increased during glucose/electrolyte absorption. By contrast, absorption of micellar oleic acid is associated with a decrease in σ_d (Figure 6.3) [225]. The σ_d for proteins of different molecular sizes has been used to calculate equivalent pore sizes in the intestinal capillaries during lipid absorption (Figure 6.4). There was no change in the size of the small pores during lipid absorption (46 vs 51 Å). However, the dimensions of the large pores increased from 200 to 300 Å. Both cholecystokinin [156,228] and neurotensin [179,229] have been shown to be released by the upper small intestine after intraluminal administration of lipids. Intravascular administration of neurotensin to achieve circulating postprandial concentrations decreased, σ_d [230] while cholecystokinin did not [225]. The neurotensin-induced increase in permeability was a result of a selective increase in the dimensions of the large pores (from 47 to 330 Å). These findings suggest that neurotensin may mediate the vascular permeability changes produced by lipid absorption. However, conclusive evidence (e.g., pharmacologic or genetic blockade) to support this contention is lacking.

Several pathologic interventions have been associated with changes in σ_d for total plasma proteins. I/R of the small intestine is associated with a decrease in σ_d for total proteins [210]. Mast cell degranulation decreases σ_d, an effect attributed to the action of histamine on H_2 receptors [231]. Experimental arterial hypertension also decreases σ_d [232]. A role for angiotensin II in this phenomenon is not likely, since angiotensin II actually increases, rather than decreases σ_d [30]. The mechanism by which angiotensin II decreases intestinal vascular permeability to plasma proteins has not been addressed, but angiotensin II decreases the movement of albumin across

TABLE 6.3: Effects of physiologic and pathologic interventions on the osmotic reflection coefficient (σ_d) for total plasma proteins in the intestinal microcirculation.

EXPERIMENTAL CONDITION	CHANGE IN σ_d
Physiologic	
Electrolyte absorption	No effect
Glucose absorption	No effect
Oleic acid absorption	Decrease (0.22)
–Cholecystokinin	No effect
–Neurotensin	Decrease (0.19)
Pathologic	
Ethanol	No effect
Mast cell degranulation	Decrease (0.16)
–Histamine	Decrease (0.36)
Ischemia/reperfusion	Decrease (0.33)
–Superoxide neutralization	Decrease (0.11)
Chronic arterial hypertension	Decrease (0.22)
–Angiotensin II	Increase (0.08)
Cirrhosis (portal hypertension)	No effect
–Glucagon	Decrease (0.10)
E. coli endotoxin	Decrease (0.14)
Miscellaneous	
Isproterenol	No effect
Bradykinin	Decrease (0.27)

Reported values for σ_d under control conditions vary from 0.78 to 0.92. Values in parentheses represent actual changes. Modified and used with permission from *Handbook of Physiology, The Gastrointestinal System I*, Chapter 39, pp. 1405–1474, 1989.

brain endothelial cell monolayers by modifying the functional activity of occludin, a tight junction protein [233]. Since angiotensin II may be one of the few endogenous vasoactive agents that actually decrease vascular permeability, further studies are warranted to more clearly define the mechanism by which angiotensin II strengthens endothelial barrier function.

6.4 ULTRASTRUCTURAL CORRELATES FOR THE FUNCTIONAL PATHWAYS

The specific capillary ultrastructural pathways used by solutes and water are a matter of debate [196, 202,204,207,234–236]. However, it is generally agreed that two different pathways must exist to explain physiological data: a large-pore pathway (175–250 Å radius) and a small-pore pathway (45–55 Å radius) [202,204,210].

The structural equivalents to the large-pore pathway are either the open fenestrae (200–300 Å radius) or the plasmalemmal vesicles/caveolae (250 Å radius) or a combination of both [199,202]. Obviously, the plasmalemmal vesicles assume a dominant role as the large-pore pathway in continuous capillaries (lacking fenestrae). Although the transcellular movement of the plasmalemmal vesicles is too slow to account for transendothelial movement of macromolecules [196,199], the transendothelial channels formed from multiple vesicles/caveolae would allow for more rapid transendothelial protein flux.

An interesting possibility that the large-pore system may be a dynamic ultrastructural entity encompassing both fenestrae and plasmalemmal vesicles has been proposed [202]. In this scheme (Figure 6.5), a plasmalemmal vesicle/caveolae (or vesicles) fuses with the endothelial cell membrane at both the blood and tissue fronts forming a transendothelial channel with two diaphragms, which eventually becomes a fenestra by a collapse to minimal length. This possibility is intriguing in that it may explain why the macromolecular permeability characteristics of the fenestrated capillaries of the gastrointestinal tract and continuous (containing only vesicles) capillaries of the hind-paw are similar (Tables 6.1 and 6.2). Structures identical to those schematically depicted in Figure 6.5C have been identified by electron microscopy and referred to as "barrel-shaped fenestrae [206]." Further, biochemical support for this possibility stems from the observation that a caveolar protein, PV-1, is also found on diaphragms of fenestrae and transendothelial channels [237]. Since current ultrastructural studies involve static approaches (a snap-shot in time) rather than continuous monitoring of structural changes, support for this contention awaits further refinement in the available technology. *In vitro* approaches may also be used to address this issue, particularly, since fenestrae have been induced in endothelial cells in culture [238,239].

The structural equivalents to the small-pore pathway in gastrointestinal capillaries are generally believed to be the diaphragmed fenestrae [42] and, perhaps, the endothelial cell junctions [205]. The junctional pathway may be more important in the continuous capillaries, such as those of the

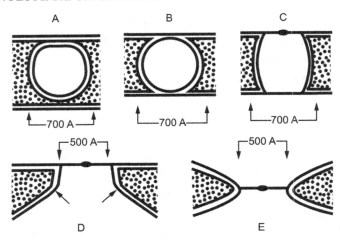

FIGURE 6.5: Proposed development of fenestrae from plasmalemmal vesicles. (A) plasmalemmal vesicle; (B) fusion of vesicle with both aspects of endothelial cell membrane; (C) formation of transendothelial channel with two diaphragms; (D and E) development of typical single diaphragmed fenestra. Used with permission from *Acta. Physiol. Scand.* 1979; Suppl. 463: pp. 11–32.

muscularis. However, the relative number of small pores predicted from functional studies (small/large pore numbers/areas >500:1; Table 6.2) do not coincide with the ratio of diaphragmed/fenestrae noted in ultrastructural studies (~50:50). The presence of a glycocalyx-like structure covering the fenestral diaphragms has been proposed to provide the necessary sieving characteristics of small pores predicted by functional studies [114,206,208].

With respect to the charge selectivity of intestinal capillaries, physiological assessments indicate that the microcirculation behaves as a positively charged barrier, thereby facilitating the transcapillary movement of negatively charged macromolecules (albumin) [222]. The presence of a glycocalyx-like structure covering the endothelial lining not only may contribute to the sieving characteristics of capillaries, but may also be an important contributor to the charge-selectivity of capillaries [114,206,208]. However, the polyanionic nature of its constituents (heparin sulfate, hyaluronic acid, etc.) imparts a negative charge to the glycocalyx. Thus, it is difficult to reconcile the positively charged barrier noted in physiological studies with the macromolecular composition of the glycocalyx. One possibility is that the composition of the glycocalyx of intestinal capillaries differs from that of glomerular capillaries. This possibility is tenable, since endothelial cells can readily alter the content and physiochemical properties of the glycocalyx [114]. However, experimental verification of this possibility is lacking.

As mentioned above, increases in microvascular permeability usually involve increases in the size of the large-pore system or the formation of capillary "gaps." Although it is generally agreed that the gaps form at the venular end of the microcirculation, the ultrastructural equivalent of these gaps is under considerable debate [240]. The formation of gaps has been attributed to contraction of endothelial cells leading to their separation [241,242]. Alternatively, it has been proposed that the plasmalemmal vesicles/caveolae (and transendothelial channels formed by their fusion) are opened wider by the endothelial cell contractile machinery [203]. Interestingly, a similar debate is ongoing regarding the ultrastructural pathway (paracellular vs vesicular) used by neutrophils to cross the venular capillaries during inflammation [243].

·   ·   ·   ·

CHAPTER 7

Transcapillary Fluid Exchange

The exchange of fluid between the blood and interstitium is dependent on the hydrostatic and colloid osmotic pressure gradients exerted across the microvasculature and by the permeability and hydraulic conductance characteristics of the capillary barrier. Accordingly, the net transcapillary fluid movement can be described by the Starling relationship

$$J_{v,c} = K_{f,c}\left[(P_c - P_t) - \sigma_d(\pi_c - \pi_t)\right]$$

where $J_{v,c}$ is the rate of net transcapillary fluid movement (capillary filtration when positive and capillary absorption when negative), $K_{f,c}$ is the capillary filtration coefficient (hydraulic conductance), P_c is the capillary hydrostatic pressure, P_t is the interstitial fluid pressure, σ_d is the osmotic reflection coefficient, π_c is the plasma oncotic pressure, and π_t is the interstitial fluid oncotic pressure.

7.1   NET TRANSCAPILLARY FLUID MOVEMENT ($J_{v,c}$)

It is generally assumed that, under resting (nontransporting) conditions, the rate of lymph flow from a tissue provides an estimate of net transcapillary filtration rate or $J_{v,c}$ [28,42,197]. Since lymph formation is dependent on interstitial to lymphatic pressure gradients, and the terminal lymphatics offer virtually no resistance to fluid movement, interstitial fluid pressure is the primary determinant of lymph flow [244]. Lymph flow is also facilitated by contractile activity of the lymphatics as well as interstitial compression of lymph vessels (e.g., villus contractions) [244,245].

The gastrointestinal tract is a transporting organ, and thus, lymph flow could vary with alterations in epithelial absorptive or secretory activity. For example, in the small intestine, net fluid absorption will increase interstitial volume (and pressure) and will drive interstitial fluid into both the capillaries and lymphatics [42,246]. Thus, the increase in lymph flow will not reflect the net transcapillary fluid movement. Conversely, net fluid secretion will result in a decrease in interstitial volume and pressure as well as an increase in capillary filtration [42]. However, a portion of the capillary filtrate will be removed from the interstitium by the secretory epithelium rather than the lymphatics. In this case, lymph flow will underestimate the net capillary filtration rate. Even if lymph flow is not affected by epithelial solute and fluid transport, as is the case in the colon [16], the rate of lymph flow will not be an accurate estimate of transcapillary filtration or absorption rate.

Under nontransporting conditions, lymph flow (in ml/min/100 g) draining the small intestine ranges from 0.005 to 0.018 in dogs [30,247], from 0.02 to 0.08 in cats [226,227,248–250], and from 0.13 to 0.38 in rats [29,251]. Thus, substantial species differences appear to exist in resting small intestinal lymph flow. A few estimates of gastric and colonic lymph flow are available. Canine colonic lymph flow is approximately 0.015 ml/min/100 g [16,218]. Direct measurements of gastric lymph flow are not available, but a rate of 0.04 ml/min/100 g has been calculated for the feline stomach from other parameters of the Starling relationship [28]. Based on the large size of the lacteals in the villi compared to the smaller, sparsely distributed initial lymphatics of the stomach and colon, one would predict that the rate of lymph flow would be greater in the small intestine than in the stomach or colon. However, the paucity of data in the stomach and colon and the degree of variability in measured resting lymph flows even within the small intestine, precludes any firm conclusions regarding a relationship between the rate of gastrointestinal lymph flow and intraorgan density/distribution of lymphatic channels.

In general, any condition or agent that increases capillary hydrostatic pressure will increase capillary fluid filtration and lymph flow [28,42]. Acute venous hypertension increases lymph flow in the stomach [212], small intestine [215], and colon [218]. Conversely, acute arterial hypotension decreases small intestinal lymph flow [60]. Under transporting conditions, changes in interstitial volume (and interstitial pressure) also produce predictable effects on small intestinal lymph flow. Net mucosal fluid absorption increases lymph flow, while net fluid secretion decreases lymph flow [28,42]. Colonic lymph flow is not affected by net mucosal fluid transport, presumably due to the remoteness of the meager initial lymphatics from the transporting epithelia [16].

7.2 CAPILLARY FILTRATION COEFFICIENT ($K_{f,c}$)

The $K_{f,c}$ is a measure of transcapillary hydraulic conductance and, as such, is influenced by both the capillary surface area available for exchange as well as the capillary permeability to solutes and fluid [27]. $K_{f,c}$ relates net transcapillary fluid movement (filtration or absorption) to the net pressure gradient established by the transcapillary hydrostatic and oncotic pressure gradients. Capillary filtration coefficients have been assessed in the stomach, small intestine, and colon using volumetric or gravimetric techniques. As mentioned above, resting values of $K_{f,c}$ range from 0.03 to 0.30 ml/min/100 g for the small intestine of cats, dogs, and rats [28–30]. A similar range of $K_{f,c}$ values have been reported for the stomach and colon [31–34].

7.3   CAPILLARY PRESSURE (P_c)

Capillary pressure, as measured by gravimetric or venous occlusion techniques, ranges from 12 to 17 mmHg in the small intestine of rats, cats, and dogs [28,42]. Capillary pressure has not been measured in gastric or colonic capillaries; however, calculated capillary pressure (from the balance of Starling forces) in the stomach is around 11 mmHg. Direct measurements of small intestinal capillary pressure by micropuncture indicate that the capillary pressure in the mucosa is lower ($\approx$15 mmHg) than that of the muscularis ($\approx$20 mmHg) [252]. The capillary pressures and relative blood flows in the two regions of the gut were used to calculate a weighted average capillary pressure of 16.8 mmHg, which agrees well with those measured with gravimetric and venous occlusion techniques.

Capillary pressure is readily altered by changes in arterial and venous pressures. Since the precapillary resistance is greater than the postcapillary resistance (pre- to postcapillary resistance ratio of 15:1), a greater increment in pressure is transmitted to the capillaries by acute venous hypertension (65–75%) than by acute arterial hypertension (5–10%) [74,250]. The increase in capillary pressure due to venous hypertension is less than predicted due to the accompanying increase in arteriolar resistance and decrease in venular resistance (increase in pre- to postcapillary resistance ratio). While this "buffering capacity" is present in the cat microcirculation, it is virtually absent in the rat, where 97% of the increment in venous pressure is transmitted back to the capillaries [29].

Vasoactive agents have predictable effects on capillary hydrostatic pressure in the gastrointestinal tract, i.e., vasodilators increase, while vasoconstrictors decrease capillary pressure. This general pattern of capillary pressure modulation by vasoactive agents is presumably due to their effect on precapillary resistance vessels [197].

7.4   INTERSTITIAL FLUID PRESSURE (P_t)

Interstitial fluid pressure in the small intestine has been assessed by (1) measurement of fluid pressure in implanted capsules, [253] (2) measurement of pressure in lacteals of the mucosal villi by micropuncture [254,255], and (3) by calculation of P_t from the balance of Starling forces [28,171,227,250]. There are limitations with all of these approaches. Implantation of capsules between the external muscle layers and the submucosa would represent submucosal/muscle interstitial fluid pressure, rather than mucosal. Lacteal pressures are assumed to equal that of the surrounding interstitial fluid due to the noncontinuous endothelial lining of the lymphatic vessels. Calculations of whole organ P_t are dependent on accurate measurements of each of the other factors of the Starling equation. Despite the different limitations of these three methods, they have yielded comparable values of P_t; under resting conditions, P_t near 0 mmHg (ranging from –2 to + 2 mmHg).

P_t is dependent on interstitial fluid volume, e.g., as gastric interstitial volume increases capsular estimates of interstitial pressure also increase [256]. The relation between intestinal interstitial fluid volume and pressure (compliance) is shown in Figure 7.1. The interstitial compliance curve indicates that, at normal tissue hydration, small changes in volume cause large changes in pressure (low compliance). However, when the tissue becomes overhydrated (edematous), a considerable volume of interstitial fluid can be accommodated with relatively small increments in pressure (high compliance). This phenomenon has been attributed to the collagen–glycosaminoglycan interactions (i.e., cross-linking), which form a compact gel-like matrix within the interstitium [28]. Thus, under resting conditions (low hydration state), this matrix immobilizes interstitial fluid such that the hydraulic conductivity of the interstitium is very low (Figure 7.1). Increases in interstitial hydration separates the matrix components (disruption of cross-linkages) and greatly increases its hydraulic conductivity [249].

Several factors have been shown to alter intestinal interstitial fluid pressure. Increases in P_t are elicited by acute venous hypertension [253], intestinal absorption [254], and local infusions of bradykinin [257], histamine [253], or glucagon [258]. Decreases in P_t are associated with acute arterial hypotension [253,259], sympathetic stimulation [248], and active secretion [260].

In the stomach, interstitial fluid pressure, as measured by capsules implanted between the submucosa and muscularis compartments [256,261], was found to be near atmospheric (0.5 ± 0.3 mmHg). Predictably, increases in venous pressure and motor activity increased interstitial fluid pressure. However, the capsule fluid pressure was not affected by transmucosal fluid secretion across the

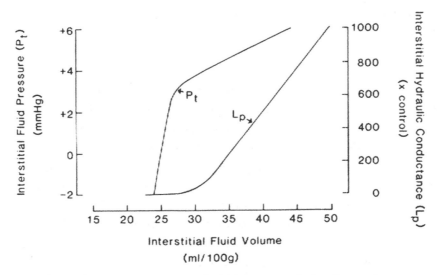

FIGURE 7.1: Relationships among interstitial fluid pressure, hydraulic conductance, and volume. Used with permission from *Handbook of Physiology, The Gastrointestinal System I*, Chapter 39, 1989, pp. 1405–1474.

gastric mucosa induced by intraluminal hypertonic solutions [261]. The fluid removed from the interstitium should have reduced mucosal interstitial fluid volume and, thereby, pressure (Figure 7.1). The lack of any changes in tissue pressure by the capsule method indicates that the submucosal/muscularis compartment (location of capsule) is situated too far from the mucosal epithelium to be impacted by transepithelial fluid movement.

7.5 OSMOTIC REFLECTION COEFFICIENT (σ_d)

Since gastrointestinal capillaries are permeable to most plasma proteins, only part of the oncotic pressure generated by plasma proteins is exerted across the capillary wall. The osmotic reflection coefficient (σ_d) describes the fraction of the total oncotic pressure generated across a capillary membrane. Impermeant proteins generate 100% of their maximum oncotic pressure ($\sigma_d = 1$), whereas freely permeable proteins do not generate an oncotic pressure ($\sigma_d = 0$). Estimates of σ_d for total plasma proteins have been obtained for all of the organ systems of the gastrointestinal tract using lymph protein flux (Table 6.1). σ_d values for total plasma proteins in the small intestine of the cat, rat, and dog range from 0.83 to 0.92 [29,30,215,230,231,247,262]. The values for σ_d in the canine colon and feline stomach are 0.085 and 0.78, respectively [212,218]. The lack of sufficient information in the stomach and colon and the variability of the estimates for σ_d in the small intestine preclude intraorgan comparisons. Various physiologic and pathologic conditions which affect σ_d are listed in Table 6.3.

7.6 TRANSCAPILLARY ONCOTIC PRESSURE GRADIENT ($\pi_c - \pi_t$)

Under the assumption that lymph protein concentration provides a valid estimate of interstitial protein concentration [216,217], the transcapillary oncotic pressure can be estimated from lymph and plasma protein concentrations using an oncometer or equations that relate protein concentration to oncotic pressure. The transcapillary oncotic pressure gradient ranges from 11.5 to 13.0 in the stomach, small intestine, and colon, with no appreciable differences among species or regions of the gastrointestinal tract [28].

Since $\sigma_d > 0$ for total plasma proteins, any condition that increases capillary filtration (in the absence of any changes in σ_d) would increase tissue volume and dilute interstitial proteins, thereby, increasing the transcapillary oncotic pressure gradient. Thus, acute venous hypertension (increased capillary pressure) increases the gradient in the stomach [212], small intestine [250], and colon [218]. Conversely, any condition that decreases capillary filtration or decrease interstitial volume would tend to decrease the transcapillary oncotic pressure gradient (e.g., arterial hypotension) [28].

In the small intestine, epithelial transport predictably effects the transcapillary oncotic pressure gradient as measured by lymphatic protein flux analysis, i.e., an increase in the gradient during fluid absorption and a decrease during fluid secretion [28,42]. However, in the colon, an analyses of lymph protein flux suggests that the transcapillary oncotic pressure gradient is not affected by either active epithelial absorption or secretion [16]. Since the lymphatic vessels are farther removed from the transporting epithelium in the colon than their counterparts in the small intestine, alterations in juxtacapillary protein concentrations may be altered and corrected (adjustments of juxtacapillary Starling forces) without concomitant detectable changes at the level of the initial lymphatics in the basal mucosal layer (crypt region) [263].

.   .   .   .

CHAPTER 8

Interaction of Capillary and Interstitial Forces

In the resting (nontransporting) gastrointestinal tract, the balance of hydrostatic and oncotic forces governing transcapillary fluid exchange favors net filtration of fluid from the blood to interstitial compartment [28,42]. To maintain a constant interstitial volume, the rate of transcapillary fluid filtration into the interstitium is balanced by an equal rate of interstitial fluid removal by the lymphatics. In this section, the interactions between capillary and interstitial forces during periods of overhydration and dehydration will be addressed. Due to the variability in the measured (or calculated) parameters of the Starling relationship and the paucity of data for the stomach and colon, emphasis will be placed on studies of the small intestine in which complete (or near complete) steady-state analyses have been reported. An example of such an analysis for the non-transporting small intestine is depicted in Figure 8.1.

8.1 INCREASED VENOUS PRESSURE

The most frequently used perturbation to assess the interaction of capillary and interstitial forces in the intestine is acute venous hypertension [42,250,264]. One such analysis is presented in Figure 8.2 [197], which depicts the adjustments in the balance of the Starling forces in response to a 14-mmHg increment in venous pressure. This maneuver increases capillary pressure by 10 mmHg, thereby producing an instantaneous increase in net filtration pressure (NFP) of 10 mmHg. This leads to enhanced transcapillary filtration of protein-poor fluid into the interstitium. As the interstitial fluid volume expands, P_t rises, and interstitial proteins are diluted, resulting in a decrease in π_t. These compensatory adjustments in interstitial forces oppose further capillary fluid filtration by reducing NFP. The increase in P_t also increases the driving pressure for lymphatic filling, and the rise in lymph flow drains some of the excess fluid from the interstitium. The net result of the interstitial compensatory adjustments is a new steady state with a slightly increased interstitial volume and an elevated lymph flow. A similar analysis in the colon has indicated that qualitatively similar compensatory adjustments in interstitial forces and lymph flow in response to an increase in capillary hydrostatic pressure occur in this region of the gastrointestinal tract [218].

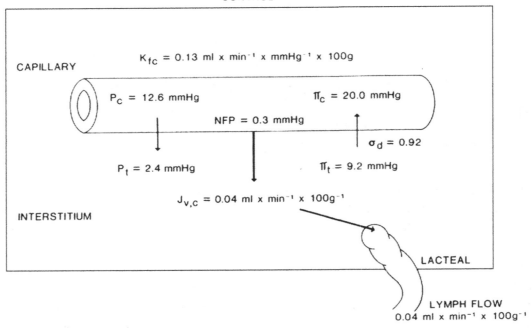

FIGURE 8.1: Starling forces and capillary membrane parameters in the small intestine under control (nontransporting) conditions. $J_{v,c}$, rate of transcapillary fluid movement; $K_{f,c}$, capillary filtration coefficient; P_c, capillary hydrostatic pressure; P_t, interstitial hydrostatic pressure; σ_d, osmotic reflection coefficient; π_c, plasma oncotic pressure; π_t, interstitial oncotic pressure; NFP, net capillary filtration pressure. Used with permission from *Handbook of Physiology, The Gastrointestinal System I*, Chapter 39, 1989, pp. 1405–1474.

Capillary exchange vessel adjustments during acute venous hypertension can modulate the efficiency of the alterations in transcapillary forces (NFP) and thereby capillary fluid filtration. As shown in Figure 3.2, the $K_{f,c}$, in the small intestine decreases with elevations in venous pressure indicating that capillary surface area is decreased. The decrease in capillary surface area serves to minimize the overall effects of the compensatory readjustments in interstitial hydrostatic and oncotic forces, thereby decreasing the net capillary filtration rate [197]. By contrast, acute venous hypertension results in an increase in $K_{f,c}$ in the colon [40], a situation which maximizes the overall effectiveness of the readjustments in interstitial forces and increases the net capillary filtration rate.

The factors of the Starling relationship that prevent excessive increases in interstitial volume in response to an increase in capillary fluid filtration have been referred to as safety factors against edema [265]. The edema safety factors are quantifiable in terms of mmHg. The relative contributions of the various edema safety factors in the feline small intestine [250] and canine colon [218]

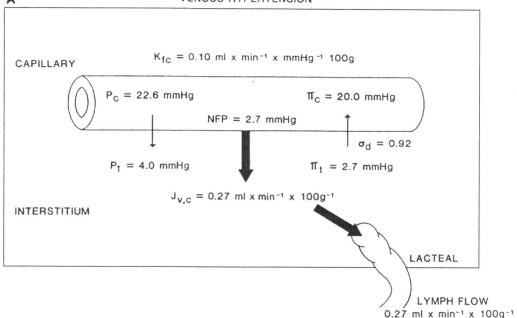

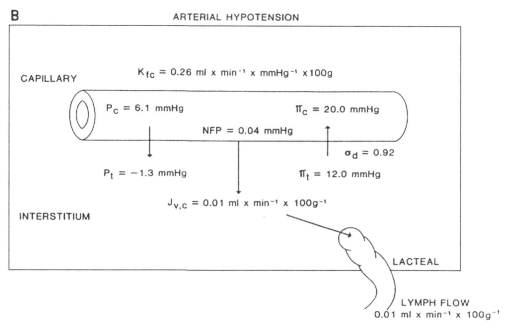

FIGURE 8.2: Effects of acute venous hypertension (A) and arterial hypotension (B) on Starling forces and capillary membrane parameters. $J_{v,c}$, rate of transcapillary fluid movement; $K_{f,c}$, capillary filtration coefficient; P_c, capillary hydrostatic pressure; P_t, interstitial hydrostatic pressure; σ_d, osmotic reflection coefficient; π_c, plasma oncotic pressure; π_t, interstitial oncotic pressure; NFP, net capillary filtration pressure. Used with permission from *Physiology of the Gastrointestinal Tract*, Second Edition, Chapter 62, 1987, pp. 1671–1697.

for an increment in capillary pressure of 12–13 mmHg are shown in Figure 8.3. In both tissues, the increased oncotic pressure gradient and interstitial fluid pressure are the major safety factors against edema, while lymph flow plays a minor role, particularly in the colon.

The edema safety factors can prevent interstitial edema until a given increment in capillary pressure is imposed such that the net transcapillary filtration rate overwhelms the safety factors. The total safety factor against edema in the small and large intestine is around 15 mmHg [42]. Increments in capillary pressure in excess of 15 mmHg lead to unopposed transcapillary fluid filtration, severe edema, and eventually exudation of fluid across the mucosa into the lumen (filtration secretion). As is the case for lymph flow, the driving force for filtration secretion is interstitial fluid pressure. For example, when villus interstitial fluid pressure is increased to 6 mmHg in the jejunum or 12 mmHg in the ileum, filtration secretion ensues [255]. Ultrastructurally, there is evidence that the mucosal epithelial barrier is compromised with separation of portions of the epithelium from the basal lamina ("bleb formation") to frank loss of epithelial cells. Thus, not only is the hydraulic conductivity of the mucosal epithelium increased, the mucosal permeability for macromolecules is

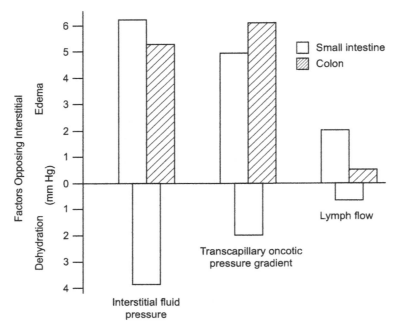

FIGURE 8.3: The safety factors against interstitial edema in the cat intestine and dog colon for an increment in capillary pressure of 12.0–13.2 mmHg as well as the safety factors against interstitial dehydration in the small intestine during a decrement in capillary pressure of 6.5 mmHg. Used with permission from *Gastroenterology* 1983; 84: pp. 846–868.

TABLE 8.1: Effects of secretagogues on intestinal lymph flow and protein concentration of secreted fluid.

SECRETAGOGUE	LYMPH FLOW × CONTROL	[PROTEIN] mg/100 ml
Acute venous hypertension	22.8	650
Histamine	20.0	1,660
Prostaglandin E_1	8.9	770
Glucagon	5.6	831
Ricinoleic acid	5.3	566
Vasoactive intestinal polypeptide	0.39	zero*
Theophylline	0.08	zero
Cholera toxin	0.03	zero

*zero refers to beyond the limits of detection. Modified and used with permission from *Handbook of Physiology, The Gastrointestinal System I*, Chapter 1, pp. 1405–1474, 1989.

increased sufficiently to allow proteins to enter the lumen (Table 8.1), i.e., the composition of the secreted fluid resembles that of lymph [263].

In addition to acute venous hypertension, a decrease in plasma oncotic pressure (π_t) by plasma dilution can induce a sufficient imbalance in forces across the capillary wall (in excess of 15 mmHg) to induce filtration secretion [42,197]. For example, acute volume expansion (normal saline) results in fluid secretion into the jejunal lumen with ultrastructural evidence of distension of the intercellular space [266]. For that matter, any condition or agents which alter the forces and/or flows governing transcapillary fluid exchange sufficiently to increase interstitial fluid pressure by 6–12 mmHg should result in filtration secretion. For example, lymph obstruction and exogenous administration of histamine and glucagon (which decrease σ_d) have been shown to produce mucosal edema, epithelial cell loss, and a secretion rich in protein (Table 8.1) [255,258,267].

Transient analyses of the relationship between filtration secretion and lymph flow indicate that as transcapillary filtration increases (acute venous hypertension or plasma dilution), lymph flow rapidly increases and then subsequently begins to fall [268]. The immediate increase in lymph flow is due to the initial increase in tissue pressure induced by enhanced capillary filtration. The subsequent decrease in lymph flow is associated with the onset of filtration secretion. A similar relationship was observed with PGE_1-induced filtration secretion, i.e., an initial progressive increase in lymph flow followed by a decrease toward basal levels with the onset of filtration secretion [267]. The decrease in lymph flow is believed to be due to the decompression of the interstitial compartment (and reduction in P_t) as a result of venting of fluid through the epithelium by filtration secretion. Under these conditions, the sum of the rates of filtration secretion and lymph flow should represent the net transcapillary filtration rate [42,197].

8.2   DECREASED ARTERIAL PRESSURE

Acute local arterial hypotension has been used to assess the interaction of capillary and interstitial forces in response to a decrease in capillary pressure (Figure 8.2). In the small intestine, a reduction in arterial pressure from 125 to 50 mmHg reduces capillary pressure by 6.5 mmHg [60]. This results in an instantaneous reversal of the net capillary filtration pressure to an absorptive pressure. The resultant movement of fluid out of the interstitium and into the capillaries causes contraction of the interstitial compartment. The reduction in interstitial volume causes a decrease in interstitial fluid pressure and an increase in interstitial oncotic pressure. These compensatory readjustments in interstitial forces not only hamper further movement of fluid out of the interstitial space, but actually result in a small net capillary filtration pressure. The decrease in interstitial fluid pressure also results in a decrease in lymph flow. In addition to adjustments in interstitial forces, perfused capillary density ($K_{f,c}$) also increases as perfusion pressure and capillary pressure is reduced (Figure 3.3). The increase in $K_{f,c}$ serves to increase the effectiveness of the small increment in NFP in preventing interstitial dehydration. Collectively, these responses result in a new steady-state transcapillary fluid balance at a slightly reduced interstitial volume, capillary filtration rate, and lymph flow.

The factors of the Starling relationship that prevent interstitial dehydration in response to a decrease in capillary fluid filtration have been referred to as safety factors against interstitial dehydration [28]. As is the case for the edema safety factors, dehydration safety factors are quantifiable in terms of mmHg. The relative contributions of the various edema safety factors in the feline small intestine for a decrement in capillary pressure of 6.5 mmHg are shown in Figure 8.3. The reduction in interstitial fluid pressure is the major factor opposing dehydration, while the increased interstitial oncotic pressure and lymph flow play more minor roles.

Stimulation of sympathetic fibers to the small intestine results in an initial vasoconstriction, which decreases blood flow to the gut. However, with prolonged stimulation, the intestinal blood vessels dilate and return toward control levels (autoregulatory escape). Thus, it was originally thought that while capillary pressure is initially decreased, it returns toward control levels during the "auto-regulatory escape" phase [42]. However, direct measurements of intestinal capillary pressure during the autoregulatory phase of sympathetic nerve stimulation indicate that it is significantly decreased. This elicits appropriate adjustments in interstitial hydrostatic and oncotic pressures to prevent interstitial dehydration [248]. The decrease in P_t and increase in π_t is qualitatively similar to those noted in response to dehydration of the interstitium induced by acute local arterial hypotension (Figure 8.2). The net effect is the attainment of new steady state with a slightly reduced interstitial volume and a decrease in lymph flow. The major qualitative difference is a reduction in $K_{f,c}$ noted with sympathetic stimulation [248] vs the increase in $K_{f,c}$ associated with acute local arterial hypertension [60]. The relative decrements in $K_{f,c}$ and lymph flow were comparable, indicating that the adjustments in the interstitial forces produced only a slight reduction in the net capillary filtration rate. Thus, the decrease in capillary surface area ($K_{f,c}$) induced by sympathetic constriction of the precapillary resistance vessels is the major factor responsible for decreasing net capillary filtration.

8.3   TRANSEPITHELIAL FLUID ABSORPTION

The major function of the small intestine is the absorption of nutrients and water. During absorption, fluid enters and expands the interstitium. In this situation, appropriate alterations in the Starling forces allow for the removal of the absorbate by the capillaries and lymphatics, thereby preventing edema.

That nutrient and fluid absorption involves alterations in intestinal transcapillary fluid exchange is evidenced by changes in various parameters of the Starling relation during ingestion of a meal or luminal placement of digested food. As mentioned above, intestinal blood flow increases when digested food is in the small intestine (postprandial hyperemia), indicating that capillary pressure (P_c) most likely increases. Further, intestinal $K_{f,c}$ has been shown to increase when digested food is in the lumen [269], indicative of an increase in capillary surface area and/or permeability. Finally, intestinal lymph flow is also increased after meals (particularly those rich in lipids) [270,271], indicating that interstitial volume and pressure are increased. The extent of interstitial hydration may be sufficient enough to actually dislodge glycosaminoglycans from the matrix and allow them to enter the systemic circulation [272]. Aside from these rather isolated observations, an extensive analysis of the alterations in the Starling parameters that occur after meals or placement of digested food in the intestinal lumen has not been undertaken. However, sufficient information is available to make some general predictions on specific alterations in the interstitial forces governing transcapillary fluid exchange during intestinal absorption using either glucose/electrolytes or fatty acids, as prototypes of hydrolytic products of food digestion.

8.3.1   Glucose and/or Electrolyte-Coupled Fluid Absorption

Although the ultimate goal of transmucosal solute and fluid absorption is the transfer of solutes and fluid from the lumen of the gut to the circulation, the first compartment impacted by glucose and/or electrolyte-coupled fluid absorption is the interstitium. As fluid enters the interstitium, the interstitial collagen–glycosaminoglycan matrix expands, resulting in several physiologic consequences that facilitate the removal of the absorbate via blood and lymph capillaries. These include (1) an increased hydraulic and solute conductivity of the matrix, (2) an increased tissue pressure, and (3) a reduced interstitial oncotic pressure [42].

In the normally hydrated interstitium, the compacted nature of the collagen–glycosaminoglycan matrix imparts a low hydraulic conductivity (Figure 7.1) [249,273]. As the interstitial volume increases during fluid absorption, the interstitial matrix expands. The expansion of the matrix is sufficient to disrupt the collagen–glycosaminoglycan interactions, and the matrix hydraulic conductivity can increase up to 1,000-fold [42,197]. This profound increase in hydraulic conductivity facilitates the movement of absorbed fluid across the interstitium to the blood and lymph microvessels. The impact of overhydration of the interstitium during absorption also has a dramatic effect on solute (macromolecule) diffusion within the matrix. For example, albumin is normally distributed in only 40% of the total matrix water due to the compact nature of the matrix. With matrix hydration and expansion during fluid absorption, albumin is more readily distributed within the interstitial matrix (up to 90%). Thus, during absorption, the convective and diffusive flux of albumin within the interstitium is facilitated.

Interstitial fluid and oncotic pressures are also affected by interstitial volume expansion during solute-coupled fluid absorption, i.e., interstitial fluid pressure (P_t) increases, and interstitial oncotic pressure (π_t) decreases [42,197,227]. During low rates of absorption, the increments in tissue pressure are greater than the decrements in interstitial oncotic pressure, presumably due to the initial low compliance of the interstitium (Figure 7.1). At higher absorption rates, the reverse holds true. Interstitial volume has increased sufficiently that it now represents a high compliance matrix, and further increases in tissue volume are associated with smaller increments in interstitial pressure but greater decrements in oncotic pressure [246].

Lymph flow also increases during net fluid absorption induced by glucose/electrolytes [42,197,227,249,274]. The contribution of lymph flow to the removal of absorbed volume is also a function of absorption rate. At low to moderate absorption rates, P_t increases dramatically due to the low interstitial compliance and drives lymph flow. At this stage, the lymphatics can contribute as much as 50% to the removal of absorbed fluid, the capillaries removing the remaining 50%. At higher absorption rates, the decrement in interstitial oncotic pressure exceeds the increment in tissue pressure, and relative contribution of lymph flow to the removal of the absorbate falls, the capillaries being the major conduits for removal of the absorbate.

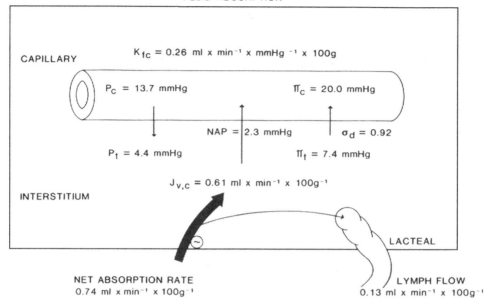

FIGURE 8.4: Effects of net fluid absorption on Starling forces and membrane parameters in the small intestine. $J_{v,c}$, rate of transcapillary fluid movement; $K_{f,c}$, capillary filtration coefficient; P_c, capillary hydrostatic pressure; P_t, interstitial hydrostatic pressure; σ_d, osmotic reflection coefficient; π_c, plasma oncotic pressure; π_t, interstitial oncotic pressure; NFP, net capillary filtration pressure. Used with permission from *Handbook of Physiology, The Gastrointestinal System I*, Chapter 39, 1989, pp. 1405–1474.

Figure 8.4 presents a steady-state analysis of the alterations in transcapillary forces and flows, which occur at a given absorption rate (0.74 ml/min/100 g) induced by perfusion of the small intestine with a 20-mM glucose solution. As fluid enters the interstitium, it begins to expand, resulting in an increase in interstitial hydrostatic pressure (P_c) of 2 mmHg and decrease in interstitial oncotic pressure (π_t) of 1.8 mmHg (compare to Figure 8.1). The microcirculatory responses to glucose absorption (vasodilation and an increase in perfused capillary density) increase capillary pressure (P_c) by 1.1 mmHg and double capillary hydraulic conductance ($K_{f,c}$). Glucose/electrolyte-coupled absorption does not alter the osmotic refection coefficient (σ_d) (Table 8.2). These alterations in capillary and interstitial forces induced by glucose absorption serve to convert the NFP to a net absorptive pressure (NAP) of 2.3 mmHg. This absorptive force, coupled to the doubling of the capillary surface area, drives 82% of the absorbate (0.61 ml/min/100 g) into the capillaries. The absorbate-induced increase in interstitial volume and, thereby, interstitial fluid pressure, increases lymph flow threefold. The enhanced lymph flow removes the remaining 18% of the absorbate

TABLE 8.2: Comparison of the alterations in the parameters of the Starling equation during glucose and oleic acid absorption.

PARAMETERS	CONTROL	GLUCOSE	OLEIC ACID
Transmucosal flow [1]	0.0	0.74	0.62
Transcapillary forces [2]			
$(P_c - P_t)$	10.5	9.3	7.3
$(\pi_t - \pi_t)$	11.0	12.6	12.3
σ_d [3]	0.92	0.92	0.70
$\sigma_d(\pi_t - \pi_t)$	10.2	11.6	8.6
$(P_c - P_t) - \sigma_d(\pi_t - \pi_t)$	+0.3	−2.3	−1.3
$K_{f,c}$ [4]	0.13	0.26	0.33
Transcapillary flow	+0.04	−0.61	−0.43
Lymphatic driving force (P_t)	2.4	4.4	7.1
Lymphatic flow	0.04	0.13	0.19

[1]All flows are in ml/min/100 gm. [2]All forces are in mm Hg. [3]The osmotic reflection coefficient (σ_d) is unitless. [4]The capillary filtration coefficient ($K_{f,c}$) is in ml/min/mm Hg/100gm. Data from Gastroenterology 1984; 86: pp. 267–273 and Am. J. Physiol. 1988; 255: pp. G690–G695.

(0.13 ml/min/100 g) from the interstitium. The net effect is the absorption of glucose-coupled fluid movement from the lumen to the circulation with a slightly increased interstitial volume.

8.3.2   Oleic Acid-Coupled Fluid Absorption

A systematic analysis of the Starling forces has also been performed during fluid absorption induced by perfusion of the small intestine with 5 mM micellar oleic acid [171]. In general, for similar rates of fluid absorption (0.62 ml/min/100 g with oleic acid vs 0.74 ml/min/100 g with glucose), the adjustments in capillary and interstitial forces induced by 5 mM oleic acid are qualitatively similar to those noted with 20 mM glucose (Table 8.2). However, there are significant quantitative differences worth noting.

Oleic acid increased interstitial fluid pressure to 7.1 mmHg, a pressure level greater than that (4.4 mmHg) noted with glucose. This greater increment in interstitial pressure indicates that the interstitial volume increased to a greater extent with oleic acid than with glucose, despite the fact that the rate of fluid absorption was greater with glucose. The significantly greater hyperemic response to oleic acid [166,171] resulted in a greater increase in capillary pressure (3 mmHg) than that noted with the glucose-induced hyperemia (1 mmHg). The net result was that the decrease in the transcapillary hydrostatic pressure gradient was greater with oleic acid than with glucose (3.2 vs 1.2 mmHg).

The transcapillary oncotic pressure gradient developed by oleic acid was similar to that noted with glucose. However, since σ_d was decreased to 0.7 by oleic acid, the *effective* transcapillary oncotic pressure gradient was actually decreased by oleic acid. This decrease in the effective oncotic pressure induced by the increase in capillary permeability to proteins decreases the ability of capillaries to absorb water and is presumed to be a major cause of the greater increase in interstitial volume observed with oleic acid.

Collectively, the adjustments in transcapillary forces were such that the NAP was only 1.3 mmHg with oleic acid in the lumen compared to 2.3 mmHg with glucose. The capillary hydraulic conductance ($K_{f,c}$) increased to a greater extent with oleic acid than with glucose (0.33 vs 0.26 ml/min/mmHg/100 g), somewhat offsetting the lower NAP. Nonetheless, 70% of the absorbate was removed by the capillaries with oleic acid compared to 82% with glucose.

Lymph flow increased approximately by threefold with glucose and fivefold with oleic acid. The greater increment in lymph flow noted with oleic acid is attributed to the higher interstitial fluid pressure incurred with oleic acid than with glucose (7.1 vs 4.4 mmHg). In addition, oleic acid increases the frequency of villus contractions, while glucose does not. [275] This increase in villus contraction frequency would also facilitate lymph flow. Thus, during oleic acid-induced fluid absorption, 30% of the absorbate was removed from the interstitium by the lymphatics, while only 18% of the absorbate was removed by the lymphatics during glucose-induced absorption.

Chylomicron Transport. The process by which fatty acids are absorbed is dependent on their chain length and water solubility. Most medium- and short-chain fatty acids are water soluble and readily absorbed by the enterocytes and enter either the capillaries or lymphatics. The absorption of the relatively water insoluble long-chain fatty acids is more complex. Long-chain fatty acids are incorporated into bile salt micelles to increase their water solubility and enhance their absorption by the enterocytes. After entering the cells, the fatty acids are re-esterified into triglycerides, provided with a glycoprotein coat, and enter the interstitium as chylomicrons [276].

Chylomicrons are large particles (400–3,000 Å radius) and cannot readily cross the capillary endothelium. Instead, chylomicrons must traverse the interstitium to reach the initial lymphatics, where they enter through the large interendothelial cell gaps [277,278] and perhaps transcellularly via vesicular transport [279]. Movement of chylomicrons through the interstitium is facilitated by the expansion of the interstitial volume during fluid absorption. The increased hydration of the interstitium disrupts the matrix structure and decreases macromolecular exclusion in the interstitial gel [249], thereby, allowing particles the size of chylomicrons to traverse the interstitium with relative ease. This scenario is supported by the observations that the rate of chylomicron transit to the lymphatics is directly related to the extent of interstitial hydration.

8.3.3 Colonic Fluid Absorption

There is a relative dearth of information regarding alterations in the Starling forces governing transcapillary fluid exchange that occur during water absorption in the colon. Direct measurements of capillary or interstitial pressure are not available. Colonic blood flow is increased by short-chain fatty acids (specifically, acetic acid), but not by an electrolyte solution [173]. Thus, it is likely that colonic capillary pressure is increased during short-chain fatty acid absorption. Lymph flow is not affected by absorption, indicating that the capillaries are the major route by which the absorbate is removed from the interstitium [16]. Lymph protein concentrations are not affected during electrolyte absorption, suggesting that the transcapillary oncotic pressure gradient is unaffected. However, the initial lymphatics of the colon sample primarily the crypt area of the mucosal interstitium, a site rather remote from the absorptive enterocytes. Thus, it cannot be assumed that juxtacapillary protein concentration (and oncotic pressure) at the epithelial surface of the mucosa has not been altered [263]. The relative location of the initial lymphatics and the lack of any changes in lymph flow during absorption preclude a systematic assessment of the Starling forces in the colon (using lymph protein flux data), as has been done for the small intestine.

8.4 TRANSEPITHELIAL FLUID SECRETION

Active fluid secretion by the gastrointestinal tract is generally accomplished by active transport of solutes (e.g., HCl or NaCl) across an intact mucosal epithelium. As opposed to the situation during solute-coupled absorption, during solute-coupled secretion, fluid is removed from the interstitium and it contracts. Adjustments in the transcapillary hydrostatic and oncotic pressure gradients ensure that fluid is removed from capillaries to prevent dehydration and provide the necessary fluid for the secretory process. The osmotically induced fluid secretion is protein-free. This is in contrast to filtration secretion (e.g., during venous hypertension), which is characterized by an increase in mucosal epithelial permeability and the passive secretion of protein-rich fluid into the lumen of the gastrointestinal tract (Table 8.1).

Information on the alterations in the Starling forces during active secretion in the stomach and colon are limited. In the stomach, pentagastrin-induced acid secretion can occur in the absence of significant changes in blood flow indicating that alterations in capillary pressure are not required for active secretion [183]. Pentagastrin does increase $K_{f,c}$ [31], but the magnitude of the increase

(50–100%) is insufficient to account for the volume of fluid that must be transferred from the micro-circulation to support the secretory process. In the colon, the only information available is the observation that induction of active secretion (theophylline) has no effect on colonic lymph flow [16].

Although a systematic assessment of the Starling forces governing transcapillary fluid exchange during active secretion has not been undertaken in the small intestine, information regarding qualitative alterations in various individual components of the Starling relationship is available. The luminal secretion induced by cholera toxin, VIP, or theophylline is devoid of protein supporting the premise that active fluid secretion occurs across an intact mucosal membrane (Table 8.1). In addition, these secretagogues decrease small intestinal lymph flow, indicative of a decrease in interstitial volume and pressure. Further, villus lacteal pressure decreases during secretion induced by cholera toxin [281], supporting the contention that interstitial volume decreases. Finally, there is evidence that cholera toxin increases blood flow (and presumably capillary pressure) and $K_{f,c}$ [282], which would favor capillary filtration. Collectively, these isolated observations, in conjunction with

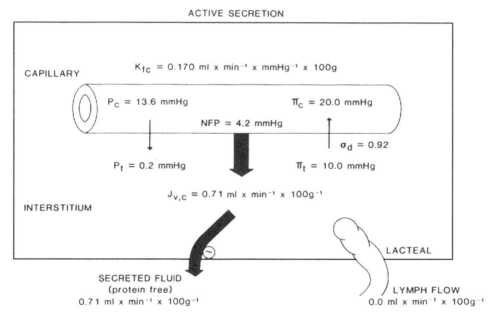

FIGURE 8.5: Effects of active (solute-coupled) fluid secretion on Starling forces and capillary membrane parameters in the small intestine. $J_{v,c}$, rate of transcapillary fluid movement; $K_{f,c}$, capillary filtration coefficient; P_c, capillary hydrostatic pressure; P_t, interstitial hydrostatic pressure; σ_d, osmotic reflection coefficient; π_c, plasma oncotic pressure; π_t, interstitial oncotic pressure; NFP, net capillary filtration pressure. Same as for Figure 8.4.

mathematical modeling approaches [283], allow for a reasonable prediction of the changes in Starling forces during active secretion in the small intestine (Figure 8.5).

The adjustments in interstitial forces during active secretion are analogous to those noted during dehydration of the interstitium induced by arterial hypotension (Figure 8.2). At the onset of active secretion, protein-free fluid is drawn from the interstitium across an intact mucosal membrane and into the lumen. Assuming a normal (low) interstitial compliance, a small decrement in interstitial volume would decrease P_t to 0.2 mmHg and increase π_t to 10 mmHg. These changes in interstitial forces, coupled to a 1.0-mmHg increment in capillary pressure (due to vasodilation) should increase net capillary filtration pressure (NFP) from 0.03 to 4.20 mmHg. The capillary filtration induced by these alterations in interstitial forces would be further enhanced by an increase in $K_{f,c}$ of approximately 30%. Since lymph flow ceases due to the fall in P_t, all of the capillary filtrate is available for the secretory process. A final confirmation of this sequence of events awaits a systematic analysis of the various parameters of the Starling relationship during active secretion under physiologic (e.g., gastrin-induced acid secretion in the stomach) or pathologic (e.g., cholera toxin-induced secretion in the small intestine) conditions.

·   ·   ·   ·

CHAPTER 9

Gastrointestinal Circulation and Mucosal Defense

Given that the primary function of the gastrointestinal tract is to assimilate nutrients from the external environment, its mucosal lining is at risk of being exposed to potentially noxious substances either ingested or endogenously produced during the course of digestion. The mucosa of the gastrointestinal tract is equipped to sense the luminal environment for potentially hazardous substances and, subsequently, mount an appropriate response to eliminate or neutralize the threat [284–286]. Upon sensing or detecting the luminal presence of potentially harmful substances, a myriad of responses is initiated by neural/humoral reflexes to deal with the threat. Alterations in gastrointestinal motor activity displace or prevent the further entrance of the noxious material [228,285]. Mucosal epithelial cell secretions (e.g., mucus, HCO_3^-) dilute and impede the access of the substances to the mucosal epithelium proper [287,288]. If the epithelial lining is breached and noxious material enters the mucosal interstitium, increases in mucosal blood flow (and associated capillary filtration) provide a means for diluting or removing the substances from the interstitium, thereby preventing more severe injury [284,287,289]. Collectively, these defensive strategies have been termed the "mucosal barrier" [290] or "mucosal defense [287,291]."

Herein, the role played by the mucosal microcirculation in the overall mucosal defense in response to exposure of the gastrointestinal mucosa to noxious substances generated postprandially will be addressed with only marginal consideration of the other components of the defense system. The most characterized reflex regulating the mucosal microcirculation during exposure of the gastrointestinal mucosa to noxious stimuli involves capsaicin-sensitive nociceptor C fibers (right side of Figure 4.1). These nerve fibers have their afferent limbs located in the mucosal interstitium and detect noxious chemicals, tactile stimuli, and/or heat via the TRPV1 receptors [133,292]. There is ample evidence to support a role for these capsaicin-sensitive nerves in the mucosal hyperemia associated with injury to the gastrointestinal tract [284–286].

9.1 GASTROINTESTINAL ACID LOAD

The average interdigestive pH of gastric lumen of humans is approximately 1.5 and can vary from 2 to 7 in the upper small intestine; cycling with intermittent gastric emptying [285,293]. The acidic environment facilitates the gastric digestion of ingested foodstuff, particularly proteins. Intragastric acid could also autodigest the stomach mucosa and, if allowed to enter in excess, the mucosa of the upper small intestine. In general, this is prevented by the mucosal defensive mechanisms mentioned above. The mucus layer covering the epithelial lining serves as a diffusional barrier for acid, and the pH at the surface of the epithelial cells is near neutral [293]. Basal H^+ secretion is associated with HCO_3^- production, which enters the gastric mucosal circulation at the base of the gastric pits and is delivered to the surface mucosal cells (Figure 2.1). In addition, the gastric and duodenal epithelial cells can generate HCO_3^-. During an enhanced acid load in either the stomach or upper small intestine, the nociceptor fibers elicit an increase in mucus production and an intense mucosal hyperemia [285,293,294]. These responses are initiated by activation of TRPV1 receptors on neurons by the presence of protons in the interstitium. The increase in mucus production serves to enhance the diffusional barrier for acid [295,296], while the mucosal hyperemia serves to neutralize and remove the acid threat from the interstitium [284,289].

The increase in mucosal blood flow during luminal acidification of the stomach is directly related to luminal $[H^+]$ [297] and appears to involve carbonic anhydrase-mediated conversion of CO_2 to carbonic acid within the epithelial cells (Figure 9.1) [289,294]. In this scenario, luminal H^+ enter the mucus gel and interact with the secreted HCO_3^- to form H_2O and CO_2. The CO_2 enters the enterocyte and is converted to carbonic acid by carbonic anhydrase. Subsequently, the carbonic acid dissociates to HCO_3^- and H^+. It is proposed that a Na^+–H^+ exchanger extrudes the H^+ into the interstitial space to activate the TRPV1 receptors on sensory neurons [284,289]. The

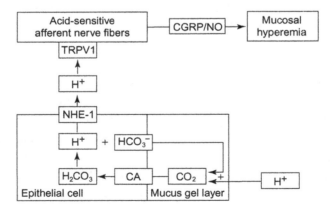

FIGURE 9.1: Proposed mechanism by which gastric intraluminal H^+ influences interstitial H^+ and can lead to activation of proton-sensitive nerves, which induce a mucosal hyperemia. CA, carbonic anhydrase; NHE-1, basolateral sodium–proton exchanger-1; CGRP, calcitonin gene-related peptide. Used with permission from *Curr. Opin. Pharmol.* 2007; 7: pp. 563–569.

increase in mucosal blood flow and mucus production is mediated by CGRP/NO; the increased mucus secretion also appears to involve a prostaglandin component [293].

The importance of the mucosal hyperemia during an acid load is underscored by the exacerbation of mucosal injury by maneuvers that limit or prevent the hyperemic response [287,291,298]. For example, even in the presence of a mucus layer, which maintains a pH gradient of near 1.0 in the lumen to 4–6 at the level of the epithelium, brief (30 s) mechanical occlusion of the arterial supply leads to dissipation of the gradient and development of deep mucosal lesions [299]. In addition, the neutrally mediated hyperemia is particularly important for neutralizing the acid threat when the mucosal defense mechanisms are compromised by various noxious agents, such as bile salts inadvertently refluxed into the stomach [298]. For example, gastric mucosal injury induced by acidified 5 mM taurocholate was associated with a mucosal hyperemia mediated by capsaicin-sensitive nerves via a CGRP/NO. The increase in mucosal blood flow prevented the deeper mucosal injury, but did not affect surface epithelial cell loss [300–302]. In addition, close intra-arterial infusion of HCO_3^- was able to protect the gastric mucosal epithelium from bile salt-induced injury [303].

9.2 INTESTINAL LIPID LOAD

Of the hydrolytic products of food digestion, only micellar oleic acid induces jejunal mucosal dysfunction/injury when placed in the lumen [305] at concentrations achievable in man [304]; hydrolytic products of carbohydrate (glucose) or protein (amino acids) digestion do not [305]. The mucosal dysfunction/injury induced by micellar oleic acid is dose-dependent. Increases in mucosal permeability (mucosal clearance of ^{51}Cr-EDTA) and filtration secretion (inhibition of absorption or conversion to frank secretion) are noted at luminal concentrations as low as 5–10 mM micellar oleic acid in pig [306,307], dog [308,305], and human [309,310]. In the dog, histologically demonstrable injury is not apparent at concentrations of 10 mM micellar oleic acid [275,308], but at 20–40 mM, the fatty acid induces loss of surface epithelial cells and subepithelial edema [305,308,311]. No substantial species differences have been reported, but the intestine of immature pigs is more sensitive to oleic acid than adult pigs [306,307]. Finally, micellar oleic acid, as well as other long-chain fatty acids, can directly injure epithelial cells (Caco-2) in culture at concentrations as low as 2–5 mM [305,308,312]. The cytotoxic effects of oleic acid on red blood cells are virtually identical to those noted in epithelial cells [312]. Taken together, it appears that oleic acid is cytotoxic to cell membranes in general.

The mechanism(s) by which micellar oleic acid induces mucosal dysfunction/injury *in vivo* is not entirely clear, but the alterations in mucosal permeability and histologically demonstrable edema and epithelial cell loss are reminiscent of "filtration secretion" [258,267]. In the cat intestine, net absorption of water and electrolytes occurs in the presence 5 mM oleic acid in the lumen

(Table 8.2) [171]. This is consistent with studies in humans where perfusion of the intestine with 5 mM oleic acid is still associated with absorption albeit at substantially reduced levels [309,310]. An analysis of the Starling forces governing transcapillary fluid exchange (Table 8.2) indicates that during lipid absorption, interstitial volume increases, and fluid leaves the interstitium via the capillaries and lymphatics. The calculated interstitial fluid pressure increases to 7.1 mmHg. In canine intestinal villi, increasing lacteal (and interstitial) pressure to approximately 9 mm Hg (range, 6.4–12.3 mm Hg) by microinjection of electrolyte solution results in shedding of epithelial cells and filtration secretion [255]. Thus, the level of interstitial fluid pressure noted during oleic acid absorption in the cat is on the cusp of that necessary to induce filtration secretion. It is conceivable that higher concentrations of oleic acid would lead to direct disruption of the epithelial lining, thereby providing for an alternate low-resistance pathway for interstitial fluid movement besides via the lymphatics and capillaries. Specifically, interstitial fluid would move across the compromised mucosal epithelium into the lumen, i.e., filtration secretion.

As is the case with mucosal dysfunction/injury, the increase in blood flow induced by micellar oleic acid is dose-dependent [172,181]. The increase in local blood flow induced by 40–80 mM micellar oleic acid can be blocked by mucosal anesthesia and capsaicin [181,313], indicating that TRPV1 receptors on sensory nerves play a role in the hyperemia. The precise mechanism(s) by which TRPV1 is activated by oleic acid in the lumen is not as well characterized as in the case of acid activation of TRPV1 in the gastric mucosa. Since the postprandial intestinal pH can fall to values of 5.0 or less [314], it may be that a sufficient amount of protons may enter the interstitium across a compromised epithelium and activate the TPRV1 receptors. Alternatively, while it is well established that TRPV1 is acid sensitive, there is evidence to indicate that it is a rather promiscuous receptor that can be activated by other stimuli, such as products of lipid metabolism [315]. One likely candidate is oleoylethanolamide, which is generated from absorbed oleic acid by intestinal epithelial cells and can activate the TRPV1 receptor [316–318]. Finally, oleic acid may stimulate intestinal epithelial cells to release substances, which then indirectly activate the TRPV1 receptor. A likely candidate is serotonin released by enterochromaffin cells in response to lipids [319,320]. Although the efferent pathway and specific mediator(s) of the hyperemia are unclear, both NO [313] and VIP [181] appear to be involved. Further work is needed to more precisely delineate the mechanism by which capsaicin-sensitive nerves are activated by luminal oleic acid and the pathway involved in the resultant hyperemia.

In addition to increasing jejunal blood flow, micellar oleic acid (10 mM) also induces mucus production [308], both of which are integral components of the mucosal defense system [287,290,291]. The increase in mucus production serves to delay both diffusion of oleic acid to the epithelial cells [308] and oleic acid-induced injury to epithelial cells *in vivo* [321] and *in vitro* [308]. Unlike the case for the stomach where the increase in blood flow contributes to the neutralization/

removal of interstitial acid, the role of blood flow in limiting/preventing the oleic acid-mediated injury is not entirely clear. Although speculative, it may be that the enhanced blood flow in combination with an increase in vascular permeability to protein serves to increase removal of oleic acid or ligands for TPRV1 (e.g., protons, oleoylethanolamide) from the interstitium via both the capillaries and lymphatics [171,225]. Interestingly, during lipid absorption by the small intestine up to twenty percent of long chain fatty acids can enter the circulation bound to albumin [276].

In the colon, acetic acid (the most prevalent short-chain fatty acid generated in the colon) does not injure the colonic mucosa [322] when administered at pH 7.4 and at postprandial concentrations, i.e., 100 mM [323]. However, if the pH of the solutions is reduced to 5.0, there is histological evidence of epithelial cell loss [322]. With further reductions in pH to 4.0 and 3.0, there is progressive subepithelial edema and lifting of entire sheets of epithelial cells. Interestingly, as was the case with oleic acid-induced mucosal injury, the developing intestine of rats is more sensitive to acidified acetic acid-induced injury than that of the adult [324]. Since the pH of the colon can be reduced to 5.0–5.5 in the human after ileal introduction of starch [325], it is conceivable that short-chain fatty acids may injure the colon postprandially.

Administration of acetic acid in the colon reinforces the colonic mucosal defense system, i.e., the short-chain fatty acid induces an increase in local blood flow [173] and mucin production [326]. The mechanisms involved are not entirely clear, but there is evidence that intracolonic acetic acid can activate TRPV1 on capsaicin-sensitive nerves via enterochromaffin cell release of serotonin [327].

9.3   GASTROINTESTINAL RESTITUTION

From a teleological perspective, it is not very appealing to envision that gastrointestinal mucosal injury is incurred after meals. The concentrations of acid, oleic acid, or short-chain fatty acids used to induce gastrointestinal mucosal injury in the experimental setting are within the levels noted after meals in humans. Nonetheless, it is quite possible that the mucosal defense system is compromised in the experimental setting, i.e., rinsing of luminal contents can remove mucus [308], and anesthesia can interfere with neural reflexes [42]. Thus, while the mucosal defense system in conscious animals or man is sufficient to minimize/prevent injury by ingested substances, in the experimental setting, it may be sufficiently compromised to be more vulnerable to a similar challenge. Nonetheless, it is generally believed that superficial damage of the gastrointestinal mucosa (epithelial cell loss) is a common occurrence and is rapidly repaired within minutes by a process termed "restitution" [328–333]. Restitution involves the migration of viable epithelial cells over the denuded basal lamina to re-establish the epithelial lining and consequently mucosal barrier function.

In the experimental setting, restitution after superficial injury has been demonstrated in the stomach [329,334,335], small intestine [305,321], and colon [322,336]. Of relevance here is that oleic acid-induced injury [305] and filtration secretion [309] in the small intestine, as well as acetic acid-induced injury [322] to the colon is reversible within minutes (restitution). That restitution may actually occur after meals in man is evidenced by the finding that after a meat meal, the gastric mucosa is characterized by the occasional presence of gaps in the lining, a mucus layer containing cells in various stages of degeneration and migrating epithelial cells [337].

• • • •

CHAPTER 10

Gastrointestinal Circulation and Mucosal Pathology I: Ischemia/Reperfusion

10.1 MODERATE REDUCTIONS IN BLOOD FLOW: DYSFUNCTION

The extramural and intramural collateral arrangement of blood vessels provides some degree of protection against ischemia during occlusion of one of the supplying arteries. For example, small intestinal blood flow is not completely compromised by occlusion of the superior mesenteric artery, since the collaterals derived from the celiac and inferior mesenteric arteries provide compensatory perfusion [338]. Species differences may exist with respect to the efficiency of the collateral circulation of the major supply vessels of the gastrointestinal tract [61]. Within the small intestine, collaterals between adjacent segments also provide protection against occlusion of a mesenteric arterial supply vessel. Occlusion of the arterial supply to one segment results in only a 45% decrease in blood flow to the segment due to the blood supply from the collaterals of the perfused adjacent segment; two thirds of this circulation is provided by extramural collaterals, while one third is derived from intramural collaterals [339]. Based on the relationship between blood flow and intestinal oxygenation (Figure 3.3), this degree of blood flow reduction is insufficient to compromise intestinal oxygenation and, therefore, function.

More severe reductions in gastrointestinal blood flow can have adverse effects on gastrointestinal function, i.e., secretory, absorptive, and motor activity. As predicted by the relationship between blood flow and oxygen uptake, gastric acid secretion is not affected by moderate reductions in blood flow. However, when blood flow is substantially reduced, such that oxygen uptake is compromised, then gastric acid output becomes dependent on blood flow and oxygen consumption [35,183,340]. Similarly, reducing intestinal blood flow by 50% does not affect glucose absorption or oxygen consumption, but further reductions in blood flow result in parallel reductions in oxygen consumption and glucose absorption [341]. Finally, moderate reductions in blood flow do not affect intestinal motility until oxygen uptake is compromised, at which point motility decreases [38]. More severe reductions in blood flow have paradoxical effects on motility [10,342]. During severe ischemia or 75% reduction in inhaled O_2, intestinal motility actually increases for several minutes, and subsequently, the gut becomes quiescent. The transient increase in motor activity can be abolished by local tetrodotoxin, indicating that it is mediated by intrinsic nerves. At the mucosal level, villus contraction frequency is unaffected by reductions in local blood flow until blood flow is reduced below the level necessary to maintain normal tissue oxygenation, at which point villus contraction frequency decreases [343].

10.2 SEVERE REDUCTIONS IN BLOOD FLOW: INJURY

Severe reductions in gastrointestinal blood flow (ischemia) can occur during various stresses, such as exercise [344], or pathologic conditions, such as arterial embolism or thrombosis [345]. In the case of physiologic stresses, reperfusion of the gastrointestinal tract occurs with cessation of the stress, while therapeutic interventions (e.g., vasodilators) are used to reperfuse the ischemic gastrointestinal tract in the clinical setting. However, reperfusion of the ischemic gastrointestinal tract results in an exacerbation of local tissue injury and, if severe enough, may lead to multiple organ dysfunction syndrome (MODS) [346]. Since the pathogenesis of ischemic injury differs from that of reperfusion injury, herein, the two will be addressed individually.

10.2.1 Ischemia-Induced Injury

In general, for a given ischemic challenge, injury to the gastrointestinal tract is more severe in the small intestine than in the colon [347,348] or stomach [349]. This enhanced vulnerability of the small intestine may be due to the unique anatomical arrangement of the mucosal microcirculation of the small intestine, compared to that of the stomach and colon. The villus arteriolar and venular capillaries run in parallel for some distance along the villus (Figure 2.2). This countercurrent arrangement of arteriole and venular segments of the villus microvasculature has prompted the proposal that O_2 can diffuse from the arteriolar compartment to the venular compartment leading to a decrease in O_2 delivery to the villus tips [350]. Direct measurements of pO_2 along the villus indicate that pO_2 at the villus tip is lower than that of the base (or shaft) of the villus [84,350]. Although this gradient of pO_2 could be due to the countercurrent exchange of O_2 within the villus, it could also be due to the higher metabolic activity of the epithelial cells of the villus tips relative to the epithelial cells at the base [351]. Further, there also appears to be a pO_2 gradient along the colonic mucosa, the pO_2 being higher at the base than at the luminal aspect [352]. Irrespective of the mechanisms responsible for the low pO_2 at the villus tips, ischemia-induced mucosal injury occurs first at the villus tips (edema and cell sloughing) and, as the ischemic episode is increased in severity, progresses downward to include the midportion and subsequently the base of the villus [347,353–357]. Ischemic injury to the colon appears to follow a similar pattern [347].

It is generally believed that hypoxia is a major contributing factor to ischemic injury [61,62]. This contention is supported by the observations that perfusion of the intestinal lumen with oxygenated, but not nitrogenated, saline can offer protection against hypotension-induced intestinal mucosal injury [358]. The mechanisms by which local tissue hypoxia results in mucosal injury are not entirely clear, but have been attributed to depletion of tissue energy stores and cellular digestion by lysosomal enzymes [61,62].

10.2.2 I/R-Induced Injury

If the ischemic insult is severe and prolonged, the injury incurred by the gastrointestinal tract is irreparable (e.g., whole wall necrosis), and surgical removal of the affected segment is warranted. If the ischemic insult is moderate, reperfusion of the ischemic region is the desired course of action. However, reperfusion of previously ischemic regions of the gastrointestinal tract usually results in an exacerbation of injury [345,346,353,356,359]. The small intestine is more vulnerable to reperfusion injury than the stomach [349] or colon [347,348]. Indeed, there may be no evidence of reperfusion-induced exacerbation of tissue injury in the stomach [349] or colon [347]. In the small intestine, the degree of reperfusion injury is dependent on the severity of the ischemic insult. Reductions in blood flow do not affect mucosal epithelial integrity after reperfusion (as assessed by mucosal clearance of albumin) until blood flow is reduced to levels where intestinal oxygen uptake is compromised by 50% or more (Figure 10.1) [360]. Since reperfusion-induced injury is a priori dependent on a previous ischemic challenge, ischemia and the associated hypoxia is generally considered as a "primer" for the reperfusion injury. Thus, most studies addressing the mechanisms involved have used experimental models of I/R. I/R-induced gastrointestinal mucosal injury has been demonstrated in the stomach [349,361], small intestine, [362,363] and colon [347,362] of animals (rodents, cats, pigs, dogs) [61,62,332] as well as man [356,357].

In addition to the epithelial cell injury incurred during I/R, the microvasculature within the lamina propria is also affected; I/R induces an increase in intestinal microvascular permeability to plasma proteins (Table 6.3) [259]. Mucosal permeability (assessed by clearance of EDTA) is not affected by ischemia, but increases upon reperfusion [364,365]. On the other hand, capillary permeability (as assessed using the osmotic reflection, σ_d) does increase during ischemia. As shown in Figure 10.2, there is a slight increase (doubling) in permeability noted during ischemia, but the bulk of the permeability increase (fourfold) is noted at reperfusion [366]. Based on the graphic analysis of σ_d for different plasma protein fractions (see Figure 6.4), I/R selectively increases the size of the large pores from 200 to 330 Å [210]. The fact that endothelial barrier dysfunction precedes mucosal barrier disruption is not entirely surprising. The endothelial lining of microvessels is the first cell type to be impacted by both ischemia (decreased oxygen delivery) and reperfusion (reintroduction

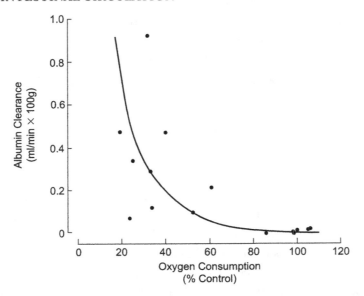

FIGURE 10.1: Relationship between ischemia/reperfusion-induced albumin clearance across the mucosa of the small intestine and the reduction in oxygen consumption induced during the ischemic period. Used with permission from *Gastroenterology* 1985; 89: pp. 852–857.

of oxygen). Since microvascular dysfunction (increased vascular permeability) is an early event in I/R, it has been the most widely used endpoint of I/R-induced intestinal injury in the experimental setting. Common approaches include measurement of the osmotic reflection coefficient (σ_d) using lymphatic protein flux in larger animals (e.g., cat) or using intravital microscopy to monitor FITC-albumin extravasation in smaller animals (e.g., mouse) [346].

10.2.3 I/R-Induced Inflammation

It is now generally accepted that the initial endothelial dysfunction (e.g., increased permeability) is a major contributor to the overall sequelae of reperfusion injury, e.g., interstitial edema and epithelial cell sloughing [62,346,367,368]. Since the pathogenesis of I/R injury is reminiscent of an inflammatory response, a great deal of effort has been devoted to the role of endothelial cells in the pathogenesis of I/R-induced inflammation. Both *in vivo* and *in vitro* approaches have been brought to bear on this issue. The *in vitro* approaches primarily involved exposure of endothelial cells to anoxia and subsequently reoxygenating (A/R) them as a simulation of *in vivo* I/R [369].

Perhaps the first significant piece of information was the observation that the I/R-induced increase in capillary permeability could be substantially blunted by SOD and implicating a role for xanthine oxidase-derived superoxide [363]. Further *in vivo* [370] and *in vitro* [371] work indicated

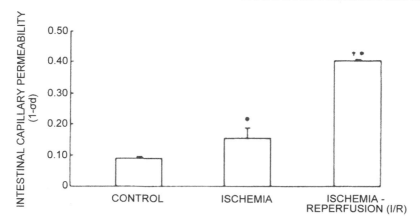

FIGURE 10.2: Effects of ischemia (blood flow reduced to 20% of control for 1 h) and ischemia/reperfusion on intestinal microvascular permeability. σ_d, osmotic reflection coefficient to total plasma proteins. *Am. J. Physiol.* 1988; 255: pp. H1269–H1275.

that the endothelium was a major source of the xanthine oxidase-derived radicals. The next step in the sequelae of I/R-induced microvascular dysfunction was the observation that xanthine oxidase-derived radicals could attract neutrophils to the intestinal mucosa [372]. Finally, it was demonstrated that infiltrating neutrophils were responsible for the I/R-induced increase in microvascular permeability [373].

The role of the endothelial cell in the inflammatory response is complex (Figure 10.3). It is generally believed that the oxidants generated during I/R (or A/R) serve as cell signaling molecules. The endothelial oxidants promote neutrophil adhesion to endothelial cells via endothelial production of chemotactic factors, such as PAF [374] and C5a [375]. A/R-induced oxidant stress also activates transcription factors involved in inflammatory gene expression (e.g., NFκB), which leads to the synthesis and surface expression of adhesion molecules on endothelial cells [376]. Another biochemical event in the endothelium after I/R is the decrease in NO bioavailability due to the interaction of NO with superoxide [377]. Since NO generated by eNOS can prevent neutrophil adherence to endothelium [378], PMN adhesion to endothelium is ensured by superoxide production after I/R [346].

However, as a caveat, it must be pointed out that virtually all of the information available on I/R-induced gastrointestinal mucosal inflammation and injury has been obtained from studies in animals. However, *in vitro* studies on endothelial cell responses to A/R have been performed using human material (HUVEC) [374,376]. These *in vitro* human models A/R accurately simulate the *in vivo* animal models of I/R and have been taken as evidence that the scenario depicted in Figure 10.3 may be applicable to humans. Further, the available studies in humans on solid organ trans-

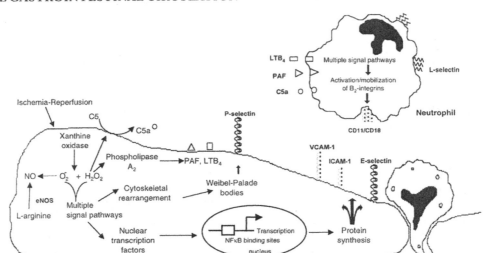

FIGURE 10.3: Proposed role of the microvascular endothelium in the inflammatory response induced by ischemia/reperfusion in the small intestine. C5a, complement 5a; CD11/CD18 and L-selectin, leukocyte adhesion glycoproteins; P-selectin, E-selectin, VCAM-1, and ICAM-1, endothelial cell adhesion glycoproteins; NFκB, nuclear factor kappa B; PAF, platelet-activating factor; LTB$_4$, leukotriene B$_4$; TNFα, tumor necrosis factor α. Used with permission from *J. Pathol.* 2000; 190: pp. 255–266.

plantation (e.g., lung, kidney) indicate that an early inflammatory component of reperfusion injury leads to eventual transplant rejection [379]. By contrast, a recent study in humans indicates that I/R injury to the small intestine is not associated with an inflammatory response [357]. The authors contend that the shedding of apoptotic and necrotic cells by the intestinal mucosa removes the stimulus for an inflammatory response, thereby protecting the intestine from inflammation-induced mucosal injury. This apparent contradiction awaits a resolution.

Collectively, the evidence obtained from *in vivo* and *in vitro* studies supports the simplified scenario depicted in Figure 10.3 [346,367,366,369,377,380]. In addition to activation of endothelial cells, I/R also activates other interstitial or circulating inflammatory cells. Activation of interstitial immune cells such as macrophages and mast cells during I/R would result in their generation of inflammatory mediators, thereby creating a chemotactic gradient to facilitate neutrophil emigration into the interstitium. A role for mast cells in I/R-induced neutrophil–endothelial cell interactions and mucosal injury has been demonstrated in the small intestine [365]. Although a direct assessment of a role for resident mucosal macrophages in I/R is difficult [381], studies in the lung show that exposure of alveolar macrophages to an oxidant stress induces them to generate a chemotactic gradient and enhance neutrophil transendothelial migration [382]. Circulating platelets also appear to modulate leukocyte–endothelial interactions after I/R [383]. Platelets, like leukocytes, adhere to endothelium after I/R [384], and roughly 50% of the leukocytes adhering to the endothelium bear attached platelets [385]. It is generally believed that the recruitment of platelets after I/R serves to exacerbate the inflammation [383].

Although I/R-induced injury is considered an innate immune response, there is a growing body of evidence to indicate that lymphocytes can modulate the inflammatory response [386]. Lymphocyte adhesion to endothelium after I/R *in vivo* [387] or A/R *in vitro* [388] occurs at a much later time point (6 h) than adhesion of neutrophils (minutes). Nonetheless, lymphocytes can modulate the early neutrophil and platelet adhesion by releasing soluble factors, such as TNF-α [389] or INFγ [390].

10.2.4 Luminal Factors May Aggravate Mucosal Injury During I/R

It is not inconceivable that after an I/R challenge, the gastrointestinal mucosal defense system would be compromised sufficiently to allow potentially harmful luminal contents to exert a detrimental effect on the mucosa. There is some evidence to indicate that I/R-induced injury to the gastrointestinal mucosa can be exacerbated by luminal acid and enzymes involved in the digestive process. Gastric mucosal injury induced by I/R can be reduced if acid secretion is pharmacologically suppressed (e.g., omeprazole); an effect prevented by exogenous administration of acid [391]. Further, inhibiting the enzymatic activity of pepsin (e.g., pepstatin) reduced the extent gastric mucosal injury induced by I/R [392]. There is also evidence that pancreatic enzymes may exacerbate I/R injury to the small intestine. Pancreatic duct ligation or luminal instillation of a broad-acting pancreatic protease inhibitor can ameliorate the mucosal injury induced by I/R [393,394]. The effects of feeding on I/R injury are somewhat equivocal. Removal of food from rats (2-day fast) provided protection from I/R-induced intestinal mucosal injury indicating that the presence of hydrolytic products of food digestion may be involved [395]. In adult pigs, luminal perfusion with digested food did not effect the I/R-induced intestinal injury [396]. Interestingly, feeding of neonatal pigs exacerbated the I/R-induced injury to the intestine; the lipid constituents of food, particularly oleic acid, were responsible [396].

The concept of intestinal "autodigestion" has been proposed to explain the role of the intestine in producing inflammatory mediators during severe mucosal injury induced in shock states that leads to multiple organ failure [397]. According to this hypothesis, during severe shock, pancreatic enzymes can digest the mucosa and liberate cytotoxic or inflammatory products, which enter the circulation and contribute to the systemic inflammation and associated sequelae. In support of the "autodigestion" hypothesis is the observation that luminal perfusion with protease and lipase inhibitors improved survival time in a splanchnic arterial occlusion model of shock [398]. It was also noted that products of lipid digestion (e.g., oleic acid) were the major cytotoxic factors.

10.3   ISCHEMIC TOLERANCE AND RESTITUTION

Cell signaling by oxidants after I/R (or A/R) can also induce a beneficial adaptive response. Specifically, challenging or preconditioning a tissue with a minor I/R insult results in protection against a major I/R challenge imposed 24 h later. This adaptive response is referred to as "ischemic tolerance," "oxidant tolerance," or "delayed preconditioning" [346,380,399]. In the small intestine, preconditioning with an I/R challenge offered protection against mucosal injury induced by a subsequent I/R challenge [364]. This adaptive protection was specific for oxidant-induced injury; preconditioning with I/R protected the intestinal mucosa from H_2O_2-induced injury, but not from that induced by acid. Ischemic tolerance was associated with an induction of antioxidant enzymes (SOD, catalase, and glutathione peroxidase) in the lamina propria [364]. The epithelial cells lining the villus and crypts did not enhance their oxidant status, nor did epithelial cells display evidence of oxidant tolerance with respect to cell injury. By contrast, endothelial cells do develop oxidant tolerance [400]. An oxidant challenge (H_2O_2) results in induction of antioxidant enzymes and results in protection from injury induced by a subsequent oxidant stress. Of particular relevance is the ability of endothelial cells to develop oxidant tolerance with respect to A/R-induced neutrophil adherence [401]. The typical increase in neutrophil adherence to endothelial cells exposed to A/R is abrogated if the endothelial cells are preconditioned 24 h earlier with an A/R challenge. This adaptive response was due to oxidant-induced activation/translocation of NFκB and induction of eNOS. *In vivo* studies indicate that I/R preconditioning prevents PMN adhesion by inhibiting the expression of P-selectin via an adenosine/PKC pathway [402]. Further work is warranted to understand the molecular mechanisms involved in ischemic preconditioning, since it is claimed to be one of the more promising strategies to ameliorate reperfusion-induced injury in the clinical setting [380].

Restitution of the intestinal mucosa after I/R-induced shedding of villus epithelial cells has been demonstrated in rats [403], dogs [362], and humans [356,357]. The time required for complete restitution varied with the duration of ischemia and degree of I/R-induced injury; the greater the injury, the longer for complete restitution [403]. As mentioned above, for a given duration of ischemia, the I/R-induced injury was more extensive in the small intestine than the colon, yet the speed of restitution was faster in the small intestine than the colon [362].

· · · · ·

CHAPTER 11

Gastrointestinal Circulation and Mucosal Pathology II: Chronic Portal Hypertension

Chronic portal hypertension (PH) is a major complication in a variety of pathologies that increase the resistance to portal inflow into the liver [404,405]. The locus of the increase in portal resistance can be prehepatic (e.g., portal vein thrombosis), intrahepatic (e.g., schistosomiasis, cirrhosis), or extrahepatic (e.g., Budd–Chiari disease) [405–407]. The most common cause of PH in Western societies is hepatic cirrhosis [408–411]. The consequences of chronic PH are not trivial, leading to the development of gastrointestinal varices and their rupture, hepatorenal syndrome, encephalopathy, and death.

Several experimental models have been developed in dogs, rats, rabbits, and mice [412–414] to study the consequences of prehepatic- and intrahepatic-initiated PH on gastrointestinal pathology. These include models of CCl₄ and bile duct ligation (cirrhosis), schistosomiasis, and simple portal vein ligation (PVL) (to produce stenosis without hepatotoxicity), the most widely used being the PVL model [412–414].

11.1 THE GASTROINTESTINAL CIRCULATION IN CHRONIC PORTAL HYPERTENSION (PH)

The evolution of compensatory alterations in the gastrointestinal vascular bed to PH has been assessed in the rat PVL model [412,415]. Paradoxically, in addition to an immediate vasoconstriction in the gastrointestinal circulation in response to an elevated portal pressure, the mesenteric vasculature becomes resistant to vasoconstrictors within 10 h after the stenosis [416]. One day after PVL, portal pressure is increased due to the increase in portal vein resistance induced by the stenosis; portal venous inflow (venous drainage from the gastrointestinal tract) is not affected nor is gastrointestinal vascular resistance [415]. By 2–4 days after PVL, portosystemic shunting progressively increases due to development of collaterals, gastrointestinal vascular resistance progressively decreases, and portal vein inflow increases. The decompression of the portal vein by developing collaterals is offset by the increase in portal inflow, and portal pressure remains elevated. Both the increased portal pressure and extent of portosystemic shunting are related to the degree of stenosis induced [417]. By 8–14 days after PVL, a steady state is achieved in which the collaterals are diverting >90% of the portal venous blood to the systemic circulation, and portal vascular resistance is normalized [415]. However, portal pressure is still elevated due to the increase in gastrointestinal blood flow and capillary pressure, which is transmitted downstream to the portal vein [415,418].

Although the PVL model has been the most widely used model of PH, it is not generally associated with liver pathology [412]. Nonetheless, the general pattern of hemodynamic events leading to PH, specifically the initial increase in portal vein resistance and subsequent increase in gastrointestinal blood flow, also hold true in experimental models of hepatic cirrhosis [413]. In the case of cirrhosis (e.g., CCl_4), the increase in portal resistance is intrahepatic and includes structural and/or functional alterations. The structural alterations such as fibrosis, sinusoidal remodeling, and vascular occlusion contribute to the increase in intrahepatic vascular resistance. In addition, due to an imbalance in vasoconstrictor and vasodilator stimuli, active constriction of stellate cells and smooth muscle cells contribute to the increase in intrahepatic vascular resistance [406,411].Collectively, the current literature supports this evolution of PH in a variety of other experimental models and the human condition [411,413], i.e., portal pressure is initially elevated due to an increase in portal resistance and subsequently maintained by the gastrointestinal hyperemia and portal inflow. The hyperemia involves all of the regions of the gastrointestinal tract, i.e., stomach, small intestine, and colon [419]. The intramural distribution of blood flow is not altered.

The increment in portal pressure achieved in various models of PH as well as the human condition is determined by both portal vascular resistance ("backward flow" mechanism) and portal inflow ("forward flow" mechanism) [404,411,413,420]. A mathematical modeling approach has provided an estimate of the relative contribution of each of the two determinants of portal pressure [420]. Figure 11.1 depicts the portal pressure–portal flow relationships generated using experimental data from normal animals (lower line) and portal hypertensive animals with established collaterals (upper line). The lines represent the vascular resistances in the normal and portal hypertensive animals. The graphic analysis indicates that either increasing portal flow in the face of a normal portal resistance (pathway A–B) or increasing vascular resistance in the absence of any change in portal inflow (pathway A–C) will result in portal hypertension. In experimental models of portal hypertension, both portal inflow and portal vascular resistance is increased (pathway A–D). Based on a 40% increase in portal resistance in experimental models of PH (0.66/0.47 in Figure 11.1), it was predicted that portal inflow accounted for 40% of the increment in portal pressure, while portal resistance accounted for the remaining 60% [62,407,420].

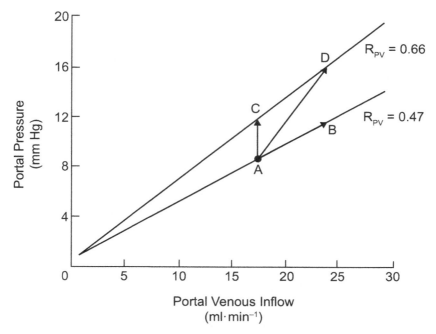

FIGURE 11.1: Relationship between portal venous pressure and portal venous inflow predicted for normal (lower line) and portal vein stenosed (upper line) rats. The effects of increasing only portal venous inflow (pathway A–B) or portal vascular resistance (pathway A–C) or a combination of both (pathway A–D) on portal pressure are shown. R_{PV} is portal vascular resistance in mmHg/ml min. Used with permission from *Pathophysiology of the Splanchnic Circulation*, Volume I, Chapter 2, 1987, pp. 57–88.

11.2 COLLATERAL VESSELS: PORTOSYSTEMIC SHUNTING

The mechanisms involved in development of collateral vascular channels to allow for portosystemic shunting (portal vein decompression) are believed to involve passive as well as active factors [404–406]. Existing portal systemic collaterals open passively in response to the increase in portal pressure. The most common sites of collateral formation are in the upper and lower portions of the gastrointestinal tract: gastro-esophageal and mesenteric–rectal varices. In addition to passive opening of preexisting channels, studies in the rat and mouse PVL model indicate that angiogenesis may contribute to the development of collaterals via a VEGF-mediated pathway [421,422]. Portosystemic shunting in this model was reduced by approximately 50% by pharmacologic or immunologic (antibody) blockade of VEGF receptor-2, indicating that a substantial amount of shunting may occur through newly formed blood vessels.

The formation of collateral channels, while serving to decompress the portal circulation, is not without detrimental consequences. An obvious consequence is that diversion of portal venous blood from the liver will interfere with hepatic uptake and metabolism of gastrointestinal-derived substances. In addition, the collateral channels formed during portal hypertension become increasingly dilated and tortuous (varices). From a clinical standpoint, the most important are the gastroesophageal varices, which can become so distended at high portal pressures that they can rupture leading to bleeding [406]. Bleeding from these varices is rather profuse and difficult to manage and can lead to death.

11.3   GASTROINTESTINAL HYPEREMIA

The PH-induced gastrointestinal hyperemia has been attributed to an imbalance in vasodilator and vasoconstrictor influences on the gastrointestinal vasculature. The two major events receiving the greatest attention are (1) hyporesponsiveness of the gastrointestinal vasculature to vasoconstrictor influences and (2) increased production and decreased hepatic metabolism of vasodilator substances.

The possibility that the gastrointestinal vasculature could be hyporesponsive to endogenous vasoconstrictors was suggested by studies in cirrhotic patients [423] and animals (dogs and rats) [424,425].The decreased mean arterial blood pressure response to angiotensin II was attributed to a postreceptor defect [425]. Studies targeting the gastrointestinal tract indicated that the sensitivity of gastric [426] and small intestinal [427] vessels to exogenous norepinephrine is reduced. Interestingly, using intravital microscopy of the small intestine and local pharmacologic antagonists, it was noted that adrenergic vasoconstrictor tone was lost in portal hypertensive animals, and the dominant vasoconstrictors were humoral, i.e., angiotensin II and vasopressin [141].

Cross-perfusion experiments provided the first indication that an increase in circulating vasoactive substances may be involved in the PH-induced gastrointestinal hyperemia. Vasodilation was induced in the intestine of naïve rats by blood derived from portal hypertensive (PVL) rats [418]. Several lines of evidence indicated that glucagon was the humoral substance responsible: (1) plasma glucagon was elevated in portal hypertensive animals, (2) glucagon induced intestinal vasodilation in naive animals when administered at concentrations measured in portal hypertensive animals, and (3) glucagon antisera attenuated the PH-induced gastrointestinal hyperemia and portal inflow by 30% [418,428]. It was proposed that enhanced production of glucagon and decreased hepatic metabolism (due to portosystemic shunting) was responsible for the glucagon-mediated gastrointestinal vasodilation in the portal hypertensive state.

There is evidence to indicate an interaction between vasoconstrictor and vasodilator systems in determining the net vascular resistance of the gastrointestinal during portal hypertension. These interactions may be paracrine or endocrine in nature. For example, the hyporesponsiveness of the mesenteric artery to α-adrenergic agonists, noted in portal hypertension, can be abrogated by either denudation of the endothelium [429] or inhibition of NO production [416]. Indeed, there is growing body of literature implicating a paracrine function of NO in the gastrointestinal hyperemia associated with portal hypertension, particularly with its role in modulating the influence of vasoconstrictors [404,413]. The only issue that has not been satisfactorily addressed is the observation that the gastrointestinal hyperemia [430] and portosystemic shunting [431] associated with portal hypertension is the same in wild-type and eNOS-deficient mice. This issue is particularly important, since eNOS-derived NO has been implicated as an initial event in the development of the gastrointestinal hyperemia in the PVL model [416] and an important contributing factor in the portal hypertension of CCl4-induced cirrhosis. [432] Circulating (endocrine) vasoactive substance, such as glucagon, have also been implicating in modulating the effects of vasoconstrictors [62]. Finally, endogenous cannabinoids (e.g., anandamide) have been implicated in gastrointestinal hyperemia in cirrhotic rats [433,434] and may be acting in a paracrine and/or endocrine fashion. Additional candidate humoral vasodilator substances implicated in the gastrointestinal vasodilation associated with portal hypertension include prostacyclin, carbon monoxide, and others [404,413].

11.4 TRANSCAPILLARY FLUID EXCHANGE

A common consequence of chronic portal hypertension is gastrointestinal edema, which generally involves the mucosal, muscle, and serosal layers [407,435]. The colon and the stomach are less susceptible to edema than the small intestine, perhaps due to efficient development of collaterals in these two regions of the gastrointestinal tract. The lymphatic vessels are dilated, and there is increase in the number of lymphatic vessels in the mesentery. Although a systematic analysis of the Starling forces has not been performed in chronic portal hypertension, some of the relevant parameters have been studied. Capillary pressure is increased due to both the increase in gastrointestinal blood flow and increase in portal pressure [418]. The osmotic reflection coefficient (σ_d) does not appear to be altered (Table 6.3), indicating that capillary permeability is not altered in cirrhotic animals. Although glucagon can increase capillary permeability (Table 6.3), this effect was noted at concentrations 1,000-fold greater than those measured in portal hypertensive animals.

In patients with advanced cirrhosis, the intestinal mucosa is characterized by villus edema and intraepithelial dilations [407]. Interestingly, portal hypertensive patients can have a portal pressure of 29 mmHg, with no evidence of filtration secretion or diarrhea [436]. In experimental animals, an acute increase in portal pressure of this magnitude would result in filtration secretion. It has been proposed that the gastrointestinal tract mucosa may adapt to chronic portal hypertension by strengthening the epithelial barrier [407,435].

11.5 LUMINAL FACTORS MAY AGGRAVATE MUCOSAL INJURY DURING PORTAL HYPERTENSION

Portal hypertension is also associated with gastropathy (portal hypertensive gastropathy; PHG) characterized by dilated mucosal blood vessels and, in severe cases, lesions indicative of intramucosal hemorrhage [437,438]. Similar vascular pathology can also be seen in the small intestine and colon [438]. PHG can result in significant bleeding, particularly in the presence of risk factors (luminal irritants such as NSAIDs, ethanol, etc.). This susceptibility to luminal irritants is believed to be due to a compromised mucosal defense system [439]. As mentioned above, two components of the mucosal defense system are mucus production and mucosal blood flow. The increase in mucus production serves to enhance the diffusional barrier for acid [295,296], while the mucosal hyperemia serves to neutralize and remove the acid threat from the interstitium [284,289]. In PH, the mucus layer is compromised in both the experimental [439,440] and clinical [441] settings. The decrease in mucus gel layer thickness in a PVL model is associated with a greater acidification of gastric epithelial cells [439]. Intraluminal acidified ethanol (15–20%) induces more injury to the gastric mucosa of cirrhotic animals than in normotensive animals; the usual gastric mucosal hyperemia is also abrogated in cirrhotic animals [440,442]. Since both the mucus and circulatory components of the gastric mucosal defense system are compromised, it is not entirely clear which defect is responsible for the susceptibility of the gastric mucosa to luminal irritants in cirrhotic animals. By contrast to the case in cirrhotic animals, the gastric mucosal injury induced by acidified ethanol is similar in control rats and portal hypertensive rats without hepatic disease (PVL) [438]. This observation suggests that hepatic dysfunction may contribute to the susceptibility of the mucosa to injury by luminal irritants. Despite the lack of a convincing body of evidence to indicate that the gastric mucosa of portal hypertensive animals is more susceptible to luminal irritants, in the clinical setting, the elimination of "risk factors" such as ethanol or NSAIDs is considered critical in the treatment of portal hypertensive gastropathy.

.   .   .   .   .

CHAPTER 12

Summary and Conclusions

In animals and man, assimilation of nutrients from the external environment is the responsibility of the gastrointestinal tract. The gastrointestinal tract is a tubular structure, which, in a strict sense, is open at both ends. Samples of the external environment are ingested orally, digested to absorbable constituents within the tract, assimilated and delivered to other organs of the body by the circulation, and the residue expelled from the distal end. The circulation of the gastrointestinal tract is critical for the overall function of this organ system, as exemplified by the fact that it receives one-fourth of the cardiac output.

The gastrointestinal circulation is responsible for supporting the enhanced functional activities associated with assimilation of nutrients, i.e., absorption, secretion, and motility. Intrinsic regulatory mechanisms modulate both resistance vessels (blood flow) and precapillary sphincters (capillary density) to ensure the adequate delivery of O_2 to meet the increased O_2 demand associated with enhanced postprandial functional activity. These regulatory mechanisms also account for the localization of hyperemia to only those regions of the small intestine through which chyme is passing. Hydrolytic products of food digestion in the lumen elicit the postprandial intestinal hyperemia, particularly long chain fatty acids (e.g., oleic acid) and monosaccharides (e.g., glucose).

The fenestrated capillaries of the gastrointestinal mucosa allow for vast amounts of transcapillary movement of solutes and fluid. Intestinal capillaries are highly permeable to small solutes, yet relatively impermeable to macromolecules (e.g., proteins). This permselectivity allows for the maintenance of a fairly constant interstitial volume during absorption by restricting proteins to the intravascular compartment, yet allowing absorbed nutrients and solutes fairly unrestricted transcapillary movement for efficient assimilation. During fluid absorption, appropriate alterations in capillary and interstitial forces allows for the movement of fluid into the capillaries and lymphatics, thereby preventing excessive hydration of the interstitium. Although the capillaries are the major conduits for fluid transport out of the interstitium, the lymphatic contribution is greater during oleic acid absorption than during glucose absorption. Although less well characterized, solute-coupled secretion is believed to also result in appropriate alterations in transcapillary forces governing fluid exchange to prevent interstitial dehydration. The water removed from the interstitium during secretion is derived solely from the capillaries at a rate sufficient to support the secretory process, thereby preventing interstitial dehydration.

The gastrointestinal circulation also contributes to the mucosal defense mechanisms whose function is to protect the mucosa from potential injury induced by noxious substances either ingested (e.g., oleic acid) or produced endogenously (e.g., acid). During an enhanced acid load in either the stomach or upper small intestine, interstitial nociceptor fibers elicit an intense mucosal hyperemia, which serves to dilute/remove the excess protons from the interstitium. A similar reflex appears to be involved in eliciting the hyperemia associated with an enhanced lipid load, albeit not as well characterized. Although the gastrointestinal circulation plays an important defensive role against noxious luminal stresses, it can also be a significant detrimental contributor to mucosal pathology elicited by stresses within the body, such as I/R and PH. The gastrointestinal microcirculation contributes to the inflammatory response elicited by I/R and the vascular congestion and edema induced by PH.

A predominant part of the database for current concepts regarding the physiology and pathophysiology of the gastrointestinal circulation is derived from studies in animals and mathematical modeling approaches. Recent technological advances (e.g., confocal endomicroscopy) should extend this database to humans and provide the necessary information for the development of rational therapeutic approaches aimed at gastrointestinal disorders involving the circulation.

·   ·   ·   ·

References

[1] Kachlik D, Baca V. Macroscopic and microscopic intermesenteric communications. 12. *Biomed Pap Med Fac Univ Palacky Olomouc Czech Repub* 2006 July;150(1): pp. 121–124.

[2] Warwick R, Williams P. *Gray's anatomy of the human body*. 36th ed. 1980. Edinburgh: Longman.

[3] Rosenblum JD, Boyle CM, Schwartz LB. The mesenteric circulation. Anatomy and Physiology. Surg Clin North Am. 1997; 77: pp. 289–306.

[4] Wheaton LG, Sarr MG, Schlossberg L, Bulkley GB. Gross anatomy of the splanchnic vasculature. In: *Measurement of Blood Flow: Applications to the Splanchnic Circulation*, Granger DN, Bulkley GB, eds. pp. 9–45. 1981. Baltimore, MD: Williams & Wilkins.

[5] Casley-Smith J, Gannon BJ. Intestinal microcirculation: spatial organization and fine structure. In: *Physiology of the Intestinal Circulation*, Shepherd A, Granger D, eds. pp. 9–31. 1984. New York: Raven Press.

[6] Sugito M, Araki K, Ogata T. Three-dimensional organization of lymphatics in the dog stomach: a scanning electron microscopic study of corrosion casts. *Arch Histol Cytol* 1996 March;59(1): pp. 61–70.

[7] Perry MA, Ardell JL, Barrowman JA, Kvietys PR. Physiology of the splanchnic circulation. In: *Pathophysiology of the Splanchnic Circulation*, Kvietys PR, Barrowman JA, Granger DN, eds. pp. 1–56. 1987. Boca Raton, FL: CRC Press.

[8] Guth PH. *In vivo* microscopy of the gastric microcirculation. In: *Measurement of Splanchnic Blood Flow: Applications to the Splanchnic Circulation*. Granger DN, Bulkley GB, eds. pp. 107–119. 1981. Baltimore, MD: Williams & Wilkins.

[9] Gallavan RH, Jr, Chou CC, Kvietys PR, Sit SP. Regional blood flow during digestion in the conscious dog. 8. *Am J Physiol* 1980 February;238(2): pp. H220–H225.

[10] Chou CC. Relationship between intestinal blood flow and motility. 405. *Annu Rev Physiol* 1982;44: pp. 29–42.

[11] Chou CC, Grassmick B. Motility and blood flow distribution within the wall of the gastrointestinal tract. 1. *Am J Physiol* 1978 July;235(1): pp. H34–H39.

[12] Gannon B, Browning J, O'Brien P, Rogers P. Mucosal microvascular architecture of the fundus and body of human stomach. *Gastroenterology* 1984 May;86(5 Pt 1): pp. 866–875.

[13] Liu H, Li YQ, Yu T, Zhao YA, Zhang JP, Zhang JN, Guo YT, Xie XJ, Zhang TG, Desmond PV. Confocal endomicroscopy for *in vivo* detection of microvascular architecture in normal and malignant lesions of upper gastrointestinal tract. 1. *J Gastroenterol Hepatol* 2008 January;23(1): pp. 56–61.

[14] Listrom MB, Fenoglio-Preiser CM. Lymphatic distribution of the stomach in normal, inflammatory, hyperplastic, and neoplastic tissue. *Gastroenterology* 1987 September;93(3): pp. 506–514.

[15] Frasher WG, Jr, Wayland H. A repeating modular organization of the microcirculation of cat mesentery. *Microvasc Res* 1972 January;4(1): pp. 62–76.

[16] Kvietys PR, Wilborn WH, Granger DN. Effects of net transmucosal volume flux on lymph flow in the canine colon. Structural–functional relationship. 1. *Gastroenterology* 1981 December;81(6): pp. 1080–1090.

[17] Benoit JN, Zawieja DC. Gastrointestinal lymphatics. In: *Physiology of the Gastrointestinal Tract*. 3rd ed. Johnson LR, ed. p. 1669. 1994. New York: Raven Press.

[18] Skinner SA, O'Brien PE. The microvascular structure of the normal colon in rats and humans. 14. *J Surg Res* 1996 March;61(2): pp. 482–490.

[19] Reynolds D, Kardon R. Methods of studying the splanchnic microvascular architecture. In: *Measurement of Blood Flow: Applications to the Splanchnic Circulation*, Granger D, Bulkley G, eds. pp. 71–88. 1981. Baltimore, MD: Williams & Wilkins.

[20] Araki K, Furuya Y, Kobayashi M, Matsuura K, Ogata T, Isozaki H. Comparison of mucosal microvasculature between the proximal and distal human colon. 1. *J Electron Microsc (Tokyo)* 1996 June;45(3): pp. 202–206.

[21] Fenoglio CM, Kaye GI, Lane N. Distribution of human colonic lymphatics in normal, hyperplastic, and adenomatous tissue. Its relationship to metastasis from small carcinomas in pedunculated adenomas, with two case reports. *Gastroenterology* 1973 January;64(1): pp. 51–66.

[22] Granger DN, Richardson PD, Kvietys PR, Mortillaro NA. Intestinal blood flow. 1. *Gastroenterology* 1980 April;78(4): pp. 837–863.

[23] Kvietys PR, Granger DN. Regulation of colonic blood flow. *Fed Proc* 1982 April;41(6): pp. 2106–2110.

[24] Lutz J, Biester J. [The reactions of the gastric vascular bed on venous or arterial pressure elevation and their comparison with values of the splenic and intestinal circulatory system. Veno-vasomotoric reaction and autoregulation]. 1. *Pflugers Arch* 1971;330(3): pp. 230–242.

[25] Granger HJ, Nyhof RA. Dynamics of intestinal oxygenation: interactions between oxygen supply and uptake. 4. *Am J Physiol* 1982 August;243(2): pp. G91–G96.

[26]   Richardson PD, Granger DN, Taylor AE. Capillary filtration coefficient: the technique and its application to the small intestine. 2. *Cardiovasc Res* 1979 October;13(10): pp. 547–561.

[27]   Granger DN, Richardson PD, Taylor AE. Volumetric assessment of the capillary filtration coefficient in the cat small intestine. 3. *Pflugers Arch* 1979 July;381(1): pp. 25–33.

[28]   Granger DN, Barrowman JA. Microcirculation of the alimentary tract I. Physiology of transcapillary fluid and solute exchange. *Gastroenterology* 1983 April;84(4): pp. 846–868.

[29]   Anzueto L, Benoit JN, Granger DN. A rat model for studying the intestinal circulation. 1. *Am J Physiol* 1984 January;246(1 Pt 1): pp. G56–G61.

[30]   Nyhof RA, Rascoe TG, Granger HJ. Acute local effects of angiotensin II on the intestinal vasculature. 1. *Hypertension* 1984 January;6(1): pp. 13–11.

[31]   Perry MA, Granger DN. Regulation of capillary exchange capacity in the dog stomach. 12. *Am J Physiol* 1985 April;248(4 Pt 1): pp. G437–G442.

[32]   Jansson G, Lundgren O, Martinson J. Neurohormonal control of gastric blood flow. 1. *Gastroenterology* 1970 March;58(3): pp. 425–429.

[33]   Fasth S, Hulten L. The effect of bradykinin on the consecutive vascular sections of the small and large intestine. 81. *Acta Chir Scand* 1973;139(8): pp. 707–715.

[34]   Richardson PD, Granger DN, Kvietys PR. Effects of norepinephrine, vasopressin, isoproterenol, and histamine on blood flow, oxygen uptake, and capillary filtration coefficient in the colon of the anesthetized dog. 3. *Gastroenterology* 1980 June;78(6): pp. 1537–1544.

[35]   Perry MA, Bulkley GB, Kvietys PR, Granger DN. Regulation of oxygen uptake in resting and pentagastrin-stimulated canine stomach. 5. *Am J Physiol* 1982 June;242(6): pp. G565–G569.

[36]   Granger DN, Kvietys PR, Perry MA. Role of exchange vessels in the regulation of intestinal oxygenation. 22. *Am J Physiol* 1982 June;242(6): pp. G570–G574.

[37]   Kvietys PR, McLendon JM, Granger DN. Postprandial intestinal hyperemia: role of bile salts in the ileum. 2. *Am J Physiol* 1981 December;241(6): pp. G469–G477.

[38]   Kvietys PR, Barrowman JA, Harper SL, Granger DN. Relations among canine intestinal motility, blood flow, and oxygenation. 1. *Am J Physiol* 1986 July;251(1 Pt 1): pp. G25–G33.

[39]   Chou CC, Nyhof RA, Kvietys PR, Sit SP, Gallavan RH, Jr. Regulation of jejunal blood flow and oxygenation during glucose and oleic acid absorption. 1. *Am J Physiol* 1985 December;249(6 Pt 1): pp. G691–G701.

[40]   Kvietys PR, Miller T, Granger DN. Intrinsic control of colonic blood flow and oxygenation. 1. *Am J Physiol* 1980 June;238(6): pp. G478–G484.

[41]   Granger DN, Perry MA, Kvietys PR, Parks DA, Benoit JN. Metabolic, myogenic, and hormonal factors in local regulation of alimentary tract blood flow. In: *Microcirculation of the Alimentary Tract.* Koo A, Lam SK, Smaje LH, eds. p. 131. 1983. Singapore: World Scientific.

[42]   Granger DN, Kvietys PR, Korthuis RJ, Premen AJ. Microcirculation of the intestinal mucosa. In: *Handbook of Physiology, The Gastrointestinal System I*. Chapter 39. pp. 1405–1474. 1989. Bethesda, MD, American Physiological Society.

[43]   Shepherd AP, Granger DN. Metabolic regulation of the intestinal circulation. In: *Physiology of the Intestinal Circulation*. Shepherd AP, Granger DN, eds. pp. 33–47. 1984. New York: Raven Press.

[44]   Granger HJ, Norris CP. Intrinsic regulation of intestinal oxygenation in the anesthetized dog. 1. *Am J Physiol* 1980 June;238(6): pp. H836–H843.

[45]   Kvietys PR, Granger DN. Physiology, pharmacology and pathology of the colonic circulation. In: *Physiology of the Intestinal Circulation*. Shepherd AP, Granger DN, eds. pp. 131–142. 1984. New York: Raven Press.

[46]   Carlson BE, Arciero JC, Secomb TW. Theoretical model of blood flow autoregulation: roles of myogenic, shear-dependent, and metabolic responses. 1. *Am J Physiol Heart Circ Physiol* 2008 October;295(4): pp. H1572–H1579.

[47]   Carlson BE, Secomb TW. A theoretical model for the myogenic response based on the length-tension characteristics of vascular smooth muscle. 3. *Microcirculation* 2005 June;12(4): pp. 327–338.

[48]   Granger DN, Granger HJ. Systems analysis of intestinal hemodynamics and oxygenation. 4. *Am J Physiol* 1983 December;245(6): pp. G786–G796.

[49]   Koenigsberger M, Sauser R, Beny JL, Meister JJ. Effects of arterial wall stress on vasomotion. 4. *Biophys J* 2006 September 1;91(5): pp. 1663–1674.

[50]   Sun D, Messina EJ, Kaley G, Koller A. Characteristics and origin of myogenic response in isolated mesenteric arterioles. 4. *Am J Physiol* 1992 November;263(5 Pt 2): pp. H1486–H1491.

[51]   Drummond HA, Grifoni SC, Jernigan NL. A new trick for an old dogma: ENaC proteins as mechanotransducers in vascular smooth muscle. 5. *Physiology (Bethesda)* 2008 February;23: pp. 23–31.

[52]   Davis MJ, Hill MA. Signaling mechanisms underlying the vascular myogenic response. 14. *Physiol Rev* 1999 April;79(2): pp. 387–423.

[53]   Johnson PC, Intaglietta M. Contributions of pressure and flow sensitivity to autoregulation in mesenteric arterioles. 30. *Am J Physiol* 1976 December;231(6): pp. 1686–1698.

[54]   Shepherd AP. Myogenic responses of intestinal resistance and exchange vessels. *Am J Physiol* 1977 November;233(5): pp. H547–H554.

[55]   Johnson PC, Hanson KM. Capillary filtration in the small intestine of the dog. 2. *Circ Res* 1966 October;19(4): pp. 766–773.

[56]   Kvietys PR, Granger DN. Effects of solute-coupled fluid absorption on blood flow and oxygen uptake in the dog colon. *Gastroenterology* 1981 September;81(3): pp. 450–457.

[57] Granger DN, Kvietys PR, Mailman D, Richardson PD. Intrinsic regulation of functional blood flow and water absorption in canine colon. 5. *J Physiol* 1980 October;307: pp. 443–451.

[58] Kuo L, Chilian WM, Davis MJ. Interaction of pressure- and flow-induced responses in porcine coronary resistance vessels. 13. *Am J Physiol* 1991 December;261(6 Pt 2): pp. H1706–H1715.

[59] Johnson PC. Myogenic and venous–arteriolar responses in the intestinal circulation. In: *Physiology of the Intestinal Circulation*. Shepherd AP, Granger DN, eds. pp. 49–60. 1984. New York: Raven Press.

[60] Granger DN, Mortillaro NA, Perry MA, Kvietys PR. Autoregulation of intestinal capillary filtration rate. 1. *Am J Physiol* 1982 December;243(6): pp. G475–G483.

[61] Crissinger KD, Granger DN. Gastrointestinal blood flow. In: *Textbook of Gastroenterology*. Yamada T, ed. 4th edition, pp. 498–520, 2003. Philadelphia: William & Wilkins.

[62] Nowicki PT, Granger DN. Gastrointestinal blood flow. In: *Textbook of Gastroenterology*. Yamada T, ed. 5th edition. pp. 540–556, 2008. Wiley-Blackwell.

[63] Kvietys PR, Perry MA, Granger DN. Intestinal capillary exchange capacity and oxygen delivery-to-demand ratio. *Am J Physiol* 1983 November;245(5 Pt 1): pp. G635–G640.

[64] Kiel JW, Riedel GL, Shepherd AP. Effects of hemodilution on gastric and intestinal oxygenation. 5. *Am J Physiol* 1989 January;256(1 Pt 2): pp. H171–H178.

[65] Holm-Rutili L, Perry MA, Granger DN. Autoregulation of gastric blood flow and oxygen uptake. 2. *Am J Physiol* 1981 August;241(2): pp. G143–G149.

[66] Kvietys PR, Granger DN. Relation between intestinal blood flow and oxygen uptake. *Am J Physiol* 1982 March;242(3): pp. G202–G208.

[67] Bulkley GB, Kvietys PR, Perry MA, Granger DN. Effects of cardiac tamponade on colonic hemodynamics and oxygen uptake. 5. *Am J Physiol* 1983 June;244(6): pp. G604–G612.

[68] Bowen JC, Garg DK, Salvato PD, Jacobson ED. Differential oxygen utilization in the stomach during vasopressin and tourniquet ischemia. 4. *J Surg Res* 1978 July;25(1): pp. 15–20.

[69] Kvietys PR, Granger DN. Vasoactive agents and splanchnic oxygen uptake. *Am J Physiol* 1982 July;243(1): pp. G1–G9.

[70] Shepherd AP. Intestinal O2 consumption and 86Rb extraction during arterial hypoxia. *Am J Physiol* 1978 March;234(3): pp. E248–E251.

[71] Shepherd AP. Intestinal capillary blood flow during metabolic hyperemia. *Am J Physiol* 1979 December;237(6): pp. E548–E554.

[72] Shepherd AP, Riedel GL. Differences in reactive hyperemia between the intestinal mucosa and muscularis. 11. *Am J Physiol* 1984 December;247(6 Pt 1): pp. G617–G622.

[73] Xu Y, Henning RH, Sandovici M, van der Want JJ, van Gilst WH, Buikema H. Enhanced myogenic constriction of mesenteric artery in heart failure relates to decreased smooth

muscle cell caveolae numbers and altered AT1- and epidermal growth factor-receptor function. 2. *Eur J Heart Fail* 2009 March;11(3): pp. 246–255.

[74] Johnson PC, Hanson KM. Effect of arterial pressure on arterial and venous resistance of intestine. *J Appl Physiol* 1962 May;17: pp. 503–508.

[75] Hanson KM, Johnson PC. Evidence for local arteriovenous reflex in intestine. *J Appl Physiol* 1962 May;17: pp. 509–513.

[76] Norris CP, Barnes GE, Smith EE, Granger HJ. Autoregulation of superior mesenteric flow in fasted and fed dogs. 1. *Am J Physiol* 1979 August;237(2): pp. H174–H177.

[77] Hanson KM, Johnson PC. Pressure–flow relationships in isolated dog colon. 1. *Am J Physiol* 1967 March;212(3): pp. 574–578.

[78] Kiel JW, Riedel GL, Shepherd AP. Local control of canine gastric mucosal blood flow. 2. *Gastroenterology* 1987 November;93(5): pp. 1041–1053.

[79] Lundgren O, Svanvik J. Mucosal hemodynamics in the small intestine of the cat during reduced perfusion pressure. 8. *Acta Physiol Scand* 1973 August;88(4): pp. 551–563.

[80] Shepherd AP. Intestinal blood flow autoregulation during foodstuff absorption. *Am J Physiol* 1980 August;239(2): pp. H156–H162.

[81] Kiel JW, Riedel GL, Shepherd AP. Autoregulation of canine gastric mucosal blood flow. 4. *Gastroenterology* 1987 July;93(1): pp. 12–20.

[82] Shepherd AP. Local control of intestinal oxygenation and blood flow. *Annu Rev Physiol* 1982;44: pp. 13–27.

[83] Bohlen HG. Intestinal mucosal oxygenation influences absorptive hyperemia. 89. *Am J Physiol* 1980 October;239(4): pp. H489–H493.

[84] Bohlen HG. Intestinal tissue pO_2 and microvascular responses during glucose exposure. 91. *Am J Physiol* 1980 February;238(2): pp. H164–H171.

[85] Lang DJ, Johnson PC. Elevated ambient oxygen does not affect autoregulation in cat mesentery. 2. *Am J Physiol* 1988 July;255(1 Pt 2): pp. H131–H137.

[86] Jacobson ED, Pawlik WW. Adenosine regulation of mesenteric vasodilation. 3. *Gastroenterology* 1994 October;107(4): pp. 1168–1180.

[87] Sawmiller DR, Chou CC. Role of adenosine in postprandial and reactive hyperemia in canine jejunum. 1. *Am J Physiol* 1992 October;263(4 Pt 1): pp. G487–G493.

[88] Sawmiller DR, Chou CC. Jejunal adenosine increases during food-induced jejunal hyperemia. 3. *Am J Physiol* 1990 March;258(3 Pt 1): pp. G370–G376.

[89] Granger HJ, Norris CP. Role of adenosine in local control of intestinal circulation in the dog. 2. *Circ Res* 1980 June;46(6): pp. 764–770.

[90] Sawmiller DR, Chou CC. Adenosine plays a role in food-induced jejunal hyperemia. 4. *Am J Physiol* 1988 August;255(2 Pt 1): pp. G168–G174.

[91]   Lautt WW. Autoregulation of superior mesenteric artery is blocked by adenosine antagonism. *Can J Physiol Pharmacol* 1986 October;64(10): pp. 1291–1295.

[92]   Pawlik WW, Hottenstein OD, Palen TE, Pawlik T, Jacobson ED. Adenosine modulates reactive hyperemia in rat gut. 2. *J Physiol Pharmacol* 1993 June;44(2): pp. 119–137.

[93]   Shepherd AP, Riedel GL, Maxwell LC, Kiel JW. Selective vasodilators redistribute intestinal blood flow and depress oxygen uptake. 2. *Am J Physiol* 1984 October;247(4 Pt 1): pp. G377–G384.

[94]   Granger DN, Valleau JD, Parker RE, Lane RS, Taylor AE. Effects of adenosine on intestinal hemodynamics, oxygen delivery, and capillary fluid exchange. 2. *Am J Physiol* 1978 December;235(6): pp. H707–H719.

[95]   Fleming I, Busse R. Molecular mechanisms involved in the regulation of the endothelial nitric oxide synthase. *Am J Physiol Regul Integr Comp Physiol* 2003 January;284(1): pp. R1–12.

[96]   Moncada S, Higgs EA. Nitric oxide and the vascular endothelium. 1. *Handb Exp Pharmacol* 2006;(176 Pt 1): pp. 213–254.

[97]   Pique JM, Esplugues JV, Whittle BJ. Endogenous nitric oxide as a mediator of gastric mucosal vasodilatation during acid secretion. 8. *Gastroenterology* 1992 January;102(1): pp. 168–174.

[98]   Bohlen HG, Lash JM. Intestinal absorption of sodium and nitric oxide-dependent vasodilation interact to dominate resting vascular resistance. 3. *Circ Res* 1996 February;78(2): pp. 231–237.

[99]   Balligand JL, Feron O, Dessy C. eNOS activation by physical forces: from short-term regulation of contraction to chronic remodeling of cardiovascular tissues. 2. *Physiol Rev* 2009 April;89(2): pp. 481–534.

[100] Bohlen HG, Nase GP. Dependence of intestinal arteriolar regulation on flow-mediated nitric oxide formation. 5. *Am J Physiol Heart Circ Physiol* 2000 November;279(5): pp. H2249–H2258.

[101] Pohl U, Busse R. Hypoxia stimulates release of endothelium-derived relaxant factor. 11. *Am J Physiol* 1989 June;256(6 Pt 2): pp. H1595–H1600.

[102] Macedo MP, Lautt WW. Autoregulatory capacity in the superior mesenteric artery is attenuated by nitric oxide. 62. *Am J Physiol* 1996 September;271(3 Pt 1): pp. G400–G404.

[103] Bohlen HG. Mechanism of increased vessel wall nitric oxide concentrations during intestinal absorption. 26. *Am J Physiol* 1998 August;275(2 Pt 2): pp. H542–H550.

[104] Steenbergen JM, Bohlen HG. Sodium hyperosmolarity of intestinal lymph causes arteriolar vasodilation in part mediated by EDRF. 2. *Am J Physiol* 1993 July;265(1 Pt 2): pp. H323–H328.

[105] Zani BG, Bohlen HG. Sodium channels are required during *in vivo* sodium chloride hyperosmolarity to stimulate increase in intestinal endothelial nitric oxide production. 2. *Am J Physiol Heart Circ Physiol* 2005 January;288(1): pp. H89–H95.

[106] Walder CE, Thiemermann C, Vane JR. Endothelium-derived relaxing factor participates in the increased blood flow in response to pentagastrin in the rat stomach mucosa. 3. *Proc Biol Sci* 1990 September 22;241(1302): pp. 195–200.

[107] Kuebler WM, Uhlig U, Goldmann T, Schael G, Kerem A, Exner K, Martin C, Vollmer E, Uhlig S. Stretch activates nitric oxide production in pulmonary vascular endothelial cells in situ. 1. *Am J Respir Crit Care Med* 2003 December 1;168(11): pp. 1391–1398.

[108] Sun D, Huang A, Recchia FA, Cui Y, Messina EJ, Koller A, Kaley G. Nitric oxide-mediated arteriolar dilation after endothelial deformation. 1. *Am J Physiol Heart Circ Physiol* 2001 February;280(2): pp. H714–H721.

[109] Awolesi MA, Sessa WC, Sumpio BE. Cyclic strain upregulates nitric oxide synthase in cultured bovine aortic endothelial cells. 2. *J Clin Invest* 1995 September;96(3): pp. 1449–1454.

[110] Segal SS. Cell-to-cell communication coordinates blood flow control. *Hypertension* 1994 June;23(6 Pt 2): pp. 1113–1120.

[111] Busse R, Fleming I. Vascular endothelium and blood flow. *Handb Exp Pharmacol* 2006; (176 Pt 2): pp. 43–78.

[112] Henrion D, Iglarz M, Levy BI. Chronic endothelin-1 improves nitric oxide-dependent flow-induced dilation in resistance arteries from normotensive and hypertensive rats. 7. *Arterioscler Thromb Vasc Biol* 1999 September;19(9): pp. 2148–2153.

[113] Pohl U, Herlan K, Huang A, Bassenge E. EDRF-mediated shear-induced dilation opposes myogenic vasoconstriction in small rabbit arteries. 1. *Am J Physiol* 1991 December;261(6 Pt 2): pp. H2016–H2023.

[114] Weinbaum S, Tarbell JM, Damiano ER. The structure and function of the endothelial glycocalyx layer. 1. *Annu Rev Biomed Eng* 2007;9: pp. 121–167.

[115] Osawa M, Masuda M, Kusano K, Fujiwara K. Evidence for a role of platelet endothelial cell adhesion molecule-1 in endothelial cell mechanosignal transduction: is it a mechanoresponsive molecule? 1. *J Cell Biol* 2002 August 19;158(4): pp. 773–785.

[116] Fleming I, Fisslthaler B, Dixit M, Busse R. Role of PECAM-1 in the shear–stress-induced activation of Akt and the endothelial nitric oxide synthase (eNOS) in endothelial cells. 18. *J Cell Sci* 2005 September 15;118(Pt 18): pp. 4103–4111.

[117] Bagi Z, Frangos JA, Yeh JC, White CR, Kaley G, Koller A. PECAM-1 mediates NO-dependent dilation of arterioles to high temporal gradients of shear stress. 24. *Arterioscler Thromb Vasc Biol* 2005 August;25(8): pp. 1590–1595.

[118]   Vanner S, Surprenant A. Neural reflexes controlling intestinal microcirculation. 1. *Am J Physiol* 1996 August;271(2 Pt 1): pp. G223–G230.

[119]   Gibbins IL, Jobling P, Morris JL. Functional organization of peripheral vasomotor pathways. 7. *Acta Physiol Scand* 2003 March;177(3): pp. 237–245.

[120]   Sjoblom-Widfeldt N. Neuro-muscular transmission in blood vessels: phasic and tonic components. An in-vitro study of mesenteric arteries of the rat. 5. *Acta Physiol Scand Suppl* 1990;587: pp. 1–52.

[121]   Greenway CV, Scott GD, Zink J. Sites of autoregulatory escape of blood flow in the mesenteric vascular bed. 5. *J Physiol* 1976 July;259(1): pp. 1–12.

[122]   Folkow B, Lewis DH, Lundgren O, Mellander S, Wallentin I. The effect of graded vasoconstrictor fibre stimulation on the intestinal resistance and capacitance vessels. 2. *Acta Physiol Scand* 1964 August;61: pp. 445–457.

[123]   Hebert MT, Marshall JM. Direct observations of responses of mesenteric microcirculation of the rat to circulating noradrenaline. 3. *J Physiol* 1985 November;368: pp. 393–407.

[124]   Shepherd AP, Riedel GL. Intramural distribution of intestinal blood flow during sympathetic stimulation. 3. *Am J Physiol* 1988 November;255(5 Pt 2): pp. H1091–H1095.

[125]   Kiel JW, Shepherd AP. Gastric oxygen uptake during autoregulatory escape from sympathetic stimulation. 3. *Am J Physiol* 1989 October;257(4 Pt 1): pp. G633–G636.

[126]   Crissinger KD, Kvietys PR, Granger DN. Autoregulatory escape from norepinephrine infusion: roles of adenosine and histamine. 3. *Am J Physiol* 1988 April;254(4 Pt 1): pp. G560–G565.

[127]   Rothe C. Control of capacitance vessels. In: *Physiology of the Intestinal Circulation*, Shepherd A, Granger D, eds. pp. 73–81. 1984. New York: Raven Press.

[128]   Vanner S. Corelease of neuropeptides from capsaicin-sensitive afferents dilates submucosal arterioles in guinea pig ileum. *Am J Physiol* 1994 October;267(4 Pt 1): pp. G650–G655.

[129]   Chen RY, Li DS, Guth PH. Role of calcitonin gene-related peptide in capsaicin-induced gastric submucosal arteriolar dilation. 1. *Am J Physiol* 1992 May;262(5 Pt 2): pp. H1350–H1355.

[130]   De FD, Wattchow DA, Costa M, Brookes SJ. Immunohistochemical characterization of the innervation of human colonic mesenteric and submucosal blood vessels. 2. *Neurogastroenterol Motil* 2008 November;20(11): pp. 1212–1226.

[131]   Holzer P. Capsaicin: cellular targets, mechanisms of action, and selectivity for thin sensory neurons. *Pharmacol Rev* 1991 June;43(2): pp. 143–201.

[132]   Chan SL, Fiscus RR. Vasorelaxations induced by calcitonin gene-related peptide, vasoactive intestinal peptide, and acetylcholine in aortic rings of endothelial and inducible nitric oxide synthase-knockout mice. 3. *J Cardiovasc Pharmacol* 2003 March;41(3): pp. 434–443.

[133] Caterina MJ, Schumacher MA, Tominaga M, Rosen TA, Levine JD, Julius D. The capsaicin receptor: a heat-activated ion channel in the pain pathway. 7. *Nature* 1997 October 23;389(6653): pp. 816–824.

[134] Scotland RS, Chauhan S, Davis C, De FC, Hunt S, Kabir J, Kotsonis P, Oh U, Ahluwalia A. Vanilloid receptor TRPV1, sensory C-fibers, and vascular autoregulation: a novel mechanism involved in myogenic constriction. 2. *Circ Res* 2004 November 12;95(10): pp. 1027–1034.

[135] Xie H, Ray PE, Short BL. Role of sensory C fibers in hypoxia/reoxygenation-impaired myogenic constriction of cerebral arteries. 1. *Neurol Res* 2009 June 30.

[136] Hottenstein OD, Pawlik WW, Remak G, Jacobson ED. Capsaicin-sensitive nerves modulate reactive hyperemia in rat gut. 4. *Proc Soc Exp Biol Med* 1992 March;199(3): pp. 311–320.

[137] Domoki F, Santha P, Bari F, Jancso G. Perineural capsaicin treatment attenuates reactive hyperaemia in the rat skin. 1. *Neurosci Lett* 2003 May 1;341(2): pp. 127–130.

[138] Remak G, Hottenstein OD, Jacobson ED. Sensory nerves mediate neurogenic escape in rat gut. *Am J Physiol* 1990 March;258(3 Pt 2): pp. H778–H786.

[139] Leung FW. Modulation of autoregulatory escape by capsaicin-sensitive afferent nerves in rat stomach. *Am J Physiol* 1992 February;262(2 Pt 2): pp. H562–H567.

[140] Vanner S, Surprenant A. Cholinergic and noncholinergic submucosal neurons dilate arterioles in guinea pig colon. 4. *Am J Physiol* 1991 July;261(1 Pt 1): pp. G136–G144.

[141] Joh T, Granger DN, Benoit JN. Endogenous vasoconstrictor tone in intestine of normal and portal hypertensive rats. 1. *Am J Physiol* 1993 January;264(1 Pt 2): pp. H171–H177.

[142] McNeill JR. Redundant nature of the vasopressin and renin–angiotensin systems in the control of mesenteric resistance vessels of the conscious fasted cat. *Can J Physiol Pharmacol* 1983 July;61(7): pp. 770–773.

[143] Shepherd AP, Pawlik W, Mailman D, Burks TF, Jacobson ED. Effects of vasoconstrictors on intestinal vascular resistance and oxygen extraction. 1. *Am J Physiol* 1976 February;230(2): pp. 298–305.

[144] Shepherd AP, Mailman D, Burks TF, Granger HJ. Effects of norepinephrine and sympathetic stimulation on extraction of oxygen and 86Rb in perfused canine small bowel. 5. *Circ Res* 1973 August;33(2): pp. 166–174.

[145] Pawlik W, Shepherd AP, Jacobson ED. Effect of vasoactive agents on intestinal oxygen consumption and blood flow in dogs. 4. *J Clin Invest* 1975 August;56(2): pp. 484–490.

[146] Pawlik WW, Shepherd AP, Mailman D, Shanbour LL, Jacobson ED. Effects of dopamine and epinephrine on intestinal blood flow and oxygen uptake. 1. *Adv Exp Med Biol* 1976;75: pp. 511–516.

[147] Folkow B, Lundgren O, Wallentin I. Studies on the relationship between flow resistance, capillary filtration coefficient and regional blood volume in the intestine of the cat. 4. *Acta Physiol Scand* 1963 March;57: pp. 270–283.

[148] Granger DN, Kvietys PR. The splanchnic circulation: intrinsic regulation. 89. *Annu Rev Physiol* 1981;43: pp. 409–418.

[149] Gallavan RH, Jr, Chou CC. Possible mechanisms for the initiation and maintenance of postprandial intestinal hyperemia. 3. *Am J Physiol* 1985 September;249(3 Pt 1): pp. G301–G308.

[150] Jeays AD, Lawford PV, Gillott R, Spencer PA, Bardhan KD, Hose DR. A framework for the modeling of gut blood flow regulation and postprandial hyperaemia. 1. *World J Gastroenterol* 2007 March 7;13(9): pp. 1393–1398.

[151] Madsen JL, Sondergaard SB, Moller S. Meal-induced changes in splanchnic blood flow and oxygen uptake in middle-aged healthy humans. 3. *Scand J Gastroenterol* 2006 January;41(1): pp. 87–92.

[152] Vatner SF, Patrick TA, Higgins CB, Franklin D. Regional circulatory adjustments to eating and digestion in conscious unrestrained primates. 20. *J Appl Physiol* 1974 May;36(5): pp. 524–529.

[153] Vatner SF, Franklin D, Van Citters RL. Mesenteric vasoactivity associated with eating and digestion in the conscious dog. 27. *Am J Physiol* 1970 July;219(1): pp. 170–174.

[154] Fronek K, Stahlgren LH. Systemic and regional hemodynamic changes during food intake and digestion in nonanesthetized dogs. 1. *Circ Res* 1968 December;23(6): pp. 687–692.

[155] Bond JH, Prentiss RA, Levitt MD. The effects of feeding on blood flow to the stomach, small bowel, and colon of the conscious dog. 13. *J Lab Clin Med* 1979 April;93(4): pp. 594–599.

[156] Fara JW, Rubinstein EH, Sonnenschein RR. Intestinal hormones in mesenteric vasodilation after intraduodenal agents. 1. *Am J Physiol* 1972 November;223(5): pp. 1058–1067.

[157] Anzueto HL, Kvietys PR, Granger DN. Postprandial hemodynamics in the conscious rat. 43. *Am J Physiol* 1986 July;251(1 Pt 1): pp. G117–G123.

[158] Skovgaard N, Conlon JM, Wang T. Evidence that neurotensin mediates postprandial intestinal hyperemia in the python, *Python regius*. 1. *Am J Physiol Regul Integr Comp Physiol* 2007 September;293(3): pp. R1393–R1399.

[159] Seth H, Sandblom E, Axelsson M. Nutrient-induced gastrointestinal hyperemia and specific dynamic action in rainbow trout (*Oncorhynchus mykiss*)—importance of proteins and lipids. 1. *Am J Physiol Regul Integr Comp Physiol* 2009 February;296(2): pp. R345–R352.

[160] Takagi T, Naruse S, Shionoya S. Postprandial celiac and superior mesenteric blood flows in conscious dogs. 3. *Am J Physiol* 1988 October;255(4 Pt 1): pp. G522–G528.

[161]   Kato M, Naruse S, Takagi T, Shionoya S. Postprandial gastric blood flow in conscious dogs. 8. *Am J Physiol* 1989 July;257(1 Pt 1): pp. G111–G117.

[162]   Brandt JL, Castleman L, Ruskin HD, Greenwald J, Kelly JJ, Jr. The effect of oral protein and glucose feeding of splanchnic blood flow and oxygen utilization in normal and cirrhotic subjects. 3. *J Clin Invest* 1955 July;34(7, Part 1): pp. 1017–1025.

[163]   Sieber C, Beglinger C, Jager K, Stalder GA. Intestinal phase of superior mesenteric artery blood flow in man. 1. *Gut* 1992 April;33(4): pp. 497–501.

[164]   Chou CC. Splanchnic and overall cardiovascular hemodynamics during eating and digestion. *Fed Proc* 1983 April;42(6): pp. 1658–1661.

[165]   Chou CC, Hsieh CP, Yu YM, Kvietys P, Yu LC, Pittman R, Dabney JM. Localization of mesenteric hyperemia during digestion in dogs. 6. *Am J Physiol* 1976 March;230(3): pp. 583–589.

[166]   Chou CC, Kvietys P, Post J, Sit SP. Constituents of chyme responsible for postprandial intestinal hyperemia. 1. *Am J Physiol* 1978 December;235(6): pp. H677–H682.

[167]   Pawlik WW, Fondacaro JD, Jacobson ED. Metabolic hyperemia in canine gut. 3. *Am J Physiol* 1980 July;239(1): pp. G12–G17.

[168]   Shepherd AP, Riedel GL. Laser-Doppler blood flowmetry of intestinal mucosal hyperemia induced by glucose and bile. 10. *Am J Physiol* 1985 April;248(4 Pt 1): pp. G393–G397.

[169]   Siregar H, Chou CC. Relative contribution of fat, protein, carbohydrate, and ethanol to intestinal hyperemia. 2. *Am J Physiol* 1982 January;242(1): pp. G27–G31.

[170]   Valleau JD, Granger DN, Taylor AE. Effect of solute-coupled volume absorption on oxygen consumption in cat ileum. 1. *Am J Physiol* 1979 February;236(2): pp. E198–E203.

[171]   Granger DN, Korthuis RJ, Kvietys PR, Tso P. Intestinal microvascular exchange during lipid absorption. 2. *Am J Physiol* 1988 November;255(5 Pt 1): pp. G690–G695.

[172]   Kvietys PR, Gallavan RH, Chou CC. Contribution of bile to postprandial intestinal hyperemia. 2. *Am J Physiol* 1980 April;238(4): pp. G284–G288.

[173]   Kvietys PR, Granger DN. Effect of volatile fatty acids on blood flow and oxygen uptake by the dog colon. 87. *Gastroenterology* 1981 May;80(5 Pt 1): pp. 962–969.

[174]   Nyhof RA, Chou CC. Evidence against local neural mechanism for intestinal postprandial hyperemia. 3. *Am J Physiol* 1983 September;245(3): pp. H437–H446.

[175]   Nyhof RA, Ingold-Wilcox D, Chou CC. Effect of atropine on digested food-induced intestinal hyperemia. 2. *Am J Physiol* 1985 December;249(6 Pt 1): pp. G685–G690.

[176]   Biber B. Vasodilator mechanisms in the small intestine. An experimental study in the cat. 131. *Acta Physiol Scand Suppl* 1973;401: pp. 1–31.

[177] Fahrenkrug J, Haglund U, Jodal M, Lundgren O, Olbe L, de Muckadell OB. Nervous release of vasoactive intestinal polypeptide in the gastrointestinal tract of cats: possible physiological implications. 1. *J Physiol* 1978 November;284: pp. 291–305.

[178] Premen AJ, Kvietys PR, Granger DN. Postprandial regulation of intestinal blood flow: role of gastrointestinal hormones. 4. *Am J Physiol* 1985 August;249(2 Pt 1): pp. G250–G255.

[179] Gallavan RH, Jr, Shaw C, Murphy RF, Buchanan KD, Joffe SN, Jacobson ED. Effects of micellar oleic acid on canine jejunal blood flow and neurotensin release. 1. *Am J Physiol* 1986 November;251(5 Pt 1): pp. G649–G655.

[180] Gallavan RH, Jr, Chen MH, Joffe SN, Jacobson ED. Vasoactive intestinal polypeptide, cholecystokinin, glucagon, and bile-oleate-induced jejunal hyperemia. 2. *Am J Physiol* 1985 February;248(2 Pt 1): pp. G208–G215.

[181] Rozsa Z, Jacobson ED. Capsaicin-sensitive nerves are involved in bile-oleate-induced intestinal hyperemia. 6. *Am J Physiol* 1989 March;256(3 Pt 1): pp. G476–G481.

[182] Baca I, Mittmann U, Feurle GE, Haas M, Muller T. Effect of neurotensin on regional intestinal blood flow in the dog. 3. *Res Exp Med (Berl)* 1981;179(1): pp. 53–58.

[183] Holm L, Perry MA. Role of blood flow in gastric acid secretion. 3. *Am J Physiol* 1988 March;254(3 Pt 1): pp. G281–G293.

[184] Cheung LY, Moody FG, Larson K, Lowry SF. Oxygen consumption during cimetidine and prostaglandin E2 inhibition of acid secretion. 2. *Am J Physiol* 1978 May;234(5): pp. E445–E450.

[185] Kowalewski K, Kolodej A. Relation between hydrogen ion secretion and oxygen consumption by ex vivo isolated canine stomach, perfused with homologous blood. 36. *Can J Physiol Pharmacol* 1972 October;50(10): pp. 955–961.

[186] Holm-Rutili L, Berglindh T. Pentagastrin and gastric mucosal blood flow. 1. *Am J Physiol* 1986 May;250(5 Pt 1): pp. G575–G580.

[187] Matheson PJ, Wilson MA, Spain DA, Harris PD, Anderson GL, Garrison RN. Glucose-induced intestinal hyperemia is mediated by nitric oxide. 11. *J Surg Res* 1997 October;72(2): pp. 146–154.

[188] Bohlen HG. Integration of intestinal structure, function, and microvascular regulation. 25. *Microcirculation* 1998;5(1): pp. 27–37.

[189] Bulbring E. Measurements of oxygen consumption in smooth muscle. *J Physiol* 1953 October;122(1): pp. 111–134.

[190] Walus KM, Fondacaro JD, Jacobson ED. Hemodynamic and metabolic changes during stimulation of ileal motility. 1. *Dig Dis Sci* 1981 December;26(12): pp. 1069–1077.

[191]   Chou CC, Gallavan RH. Blood flow and intestinal motility. 5. *Fed Proc* 1982 April;41(6): pp. 2090–2095.

[192]   Fioramonti J, Bueno L. Relation between intestinal motility and mesenteric blood flow in the conscious dog. *Am J Physiol* 1984 February;246(2 Pt 1): pp. G108–G113.

[193]   Sidky M, Bean JW. Influence of rhythmic and tonic contraction of intestinal muscle on blood flow and blood reservoir capacity in dog intestine. 2. *Am J Physiol* 1958 May;193(2): pp. 386–392.

[194]   Duncker DJ, Bache RJ. Regulation of coronary blood flow during exercise. 1. *Physiol Rev* 2008 July;88(3): pp. 1009–1086.

[195]   Segal SS. Regulation of blood flow in the microcirculation. 18. *Microcirculation* 2005 January;12(1): pp. 33–45.

[196]   Renkin EM. Multiple pathways of capillary permeability. 44. *Circ Res* 1977 December;41(6): pp. 735–743.

[197]   Granger DN, Kvietys PR, Perry MA, Barrowman JA. The microcirculation and intestinal transport In: *Physiology of the Gastrointestinal Tract*. 2nd ed. Johnson LR, ed. pp. 1671–1697. 1987. New York: Raven Press.

[198]   Casley-Smith JR, Gannon BJ. Intestinal microcirculation: spatial organization and fine structure. In: *Physiology of the Intestinal Circulation*. Shepherd AP, Granger DN. pp. 9–31. 1984. New York: Raven Press.

[199]   Simionescu N, Simionescu M, Palade GE. Permeability of intestinal capillaries. Pathway followed by dextrans and glycogens. 21. *J Cell Biol* 1972 May;53(2): pp. 365–392.

[200]   Casley-Smith JR. Endothelial fenestrae in intestinal villi: differences between the arterial and venous ends of the capillaries. 135. *Microvasc Res* 1971 January;3(1): pp. 49–68.

[201]   Clementi F, Palade GE. Intestinal capillaries. I. Permeability to peroxidase and ferritin. 2. *J Cell Biol* 1969 April;41(1): pp. 33–58.

[202]   Palade GE, Simionescu M, Simionescu N. Structural aspects of the permeability of the microvascular endothelium. 10. *Acta Physiol Scand Suppl* 1979;463: pp. 11–32.

[203]   Feng D, Nagy JA, Pyne K, Hammel I, Dvorak HF, Dvorak AM. Pathways of macromolecular extravasation across microvascular endothelium in response to VPF/VEGF and other vasoactive mediators. 3. *Microcirculation* 1999 March;6(1): pp. 23–44.

[204]   Rippe B, Haraldsson B. Transport of macromolecules across microvascular walls: the two-pore theory. 4. *Physiol Rev* 1994 January;74(1): pp. 163–219.

[205]   Michel CC, Curry FE. Microvascular permeability. 1. *Physiol Rev* 1999 July;79(3): pp. 703–761.

[206]   Rostgaard J, Qvortrup K. Electron microscopic demonstrations of filamentous molecular sieve plugs in capillary fenestrae. 5. *Microvasc Res* 1997 January;53(1): pp. 1–13.

[207] Curry FR. Microvascular solute and water transport. 14. *Microcirculation* 2005 January; 12(1): pp. 17–31.

[208] Salmon AH, Neal CR, Sage LM, Glass CA, Harper SJ, Bates DO. Angiopoietin-1 alters microvascular permeability coefficients *in vivo* via modification of endothelial glycocalyx. 2. *Cardiovasc Res* 2009 July 1;83(1): pp. 24–33.

[209] Palade GE, Bruns RR. Structural modulations of plasmalemmal vesicles. 1. *J Cell Biol* 1968 June;37(3): pp. 633–649.

[210] Taylor A, Granger D. Exchange of macromolecules across the microcirculation. In: *Handbook of Physiology. The Cardiovascular System. Microcirculation.* Renkin EM, Michel CC. Section 2, Volume IV, Part 1, Chapter 11. pp. 467–520. 1984. Washington, DC, American Physiological Society.

[211] Perry M, Granger D. Permeability characteristics of intestinal capillaries. *Physiology of the Intestinal Circulation.* Shepherd AP, Granger DN, eds. pp. 233–248. 1984. New York: Raven Press.

[212] Perry MA, Crook WJ, Granger DN. Permeability of gastric capillaries to small and large molecules. 35. *Am J Physiol* 1981 December;241(6): pp. G478–G486.

[213] Perry MA, Granger DN. Permeability of intestinal capillaries to small molecules. 39. *Am J Physiol* 1981 July;241(1): pp. G24–G30.

[214] Renkin EM. Filtration, diffusion, and molecular sieving through porous cellulose membranes. 85. *J Gen Physiol* 1954 November 20;38(2): pp. 225–243.

[215] Granger DN, Taylor AE. Permeability of intestinal capillaries to endogenous macromolecules. 23. *Am J Physiol* 1980 April;238(4): pp. H457–H464.

[216] Rutili G, Arfors KE. Protein concentration in interstitial and lymphatic fluids from the subcutaneous tissue. 6. *Acta Physiol Scand* 1977 January;99(1): pp. 1–8.

[217] Zawieja DC, Barber BJ. Lymph protein concentration in initial and collecting lymphatics of the rat. 3. *Am J Physiol* 1987 May;252(5 Pt 1): pp. G602–G606.

[218] Richardson PD, Granger DN, Mailman D, Kvietys PR. Permeability characteristics of colonic capillaries. 1. *Am J Physiol* 1980 October;239(4): pp. G300–G305.

[219] Perry MA, Navia CA, Granger DN, Parker JC, Taylor AE. Calculation of equivalent pore radii in dog hindpaw capillaries using endogenous lymph and plasma proteins. 1. *Microvasc Res* 1983 September;26(2): pp. 250–253.

[220] Renkin EM, Watson PD, Sloop CH, Joyner WM, Curry FE. Transport pathways for fluid and large molecules in microvascular endothelium of the dog's paw. 1. *Microvasc Res* 1977 September;14(2): pp. 205–214.

[221] Perry MA, Shepherd AP, Kvietys PR, Granger DN. Effect of hypoxia on feline intestinal capillary permeability. 1. *Am J Physiol* 1985 March;248(3 Pt 1): pp. G272–G276.

[222] Perry MA, Benoit JN, Kvietys PR, Granger DN. Restricted transport of cationic macromolecules across intestinal capillaries. 1. *Am J Physiol* 1983 October;245(4): pp. G568–G572.

[223] Baldwin AL, Wilson LM. Stationary red blood cells induce a negative charge on mucosal capillary endothelium. *Am J Physiol* 1994 April;266(4 Pt 1): pp. G685–G694.

[224] Deen WM, Lazzara MJ, Myers BD. Structural determinants of glomerular permeability. 5. *Am J Physiol Renal Physiol* 2001 October;281(4): pp. F579–F596.

[225] Granger DN, Perry MA, Kvietys PR, Taylor AE. Permeability of intestinal capillaries: effects of fat absorption and gastrointestinal hormones. 6. *Am J Physiol* 1982 March;242(3): pp. G194–G201.

[226] Granger DN, Taylor AE. Effects of solute-coupled transport on lymph flow and oncotic pressures in cat ileum. 38. *Am J Physiol* 1978 October;235(4): pp. E429–E436.

[227] Granger DN, Perry MA, Kvietys PR, Taylor AE. Capillary and interstitial forces during fluid absorption in the cat small intestine. 9. *Gastroenterology* 1984 February;86(2): pp. 267–273.

[228] Holzer HH, Turkelson CM, Solomon TE, Raybould HE. Intestinal lipid inhibits gastric emptying via CCK and a vagal capsaicin-sensitive afferent pathway in rats. 1. *Am J Physiol* 1994 October;267(4 Pt 1): pp. G625–G629.

[229] Fujimura M, Khalil T, Sakamoto T, Greeley GH, Jr, Salter MG, Townsend CM, Jr, Thompson JC. Release of neurotensin by selective perfusion of the jejunum with oleic acid in dogs. 1. *Gastroenterology* 1989 June;96(6): pp. 1502–1505.

[230] Harper SL, Barrowman JA, Kvietys PR, Granger DN. Effect of neurotensin on intestinal capillary permeability and blood flow. 8. *Am J Physiol* 1984 August;247(2 Pt 1): pp. G161–G166.

[231] Mortillaro NA, Granger DN, Kvietys PR, Rutili G, Taylor AE. Effects of histamine and histamine antagonists on intestinal capillary permeability. 2. *Am J Physiol* 1981 May;240(5): pp. G381–G386.

[232] Laine GA, Granger HJ. Permeability of intestinal microvessels in chronic arterial hypertension. 4. *Hypertension* 1983 September;5(5): pp. 722–727.

[233] Wosik K, Cayrol R, Dodelet-Devillers A, Berthelet F, Bernard M, Moumdjian R, Bouthillier A, Reudelhuber TL, Prat A. Angiotensin II controls occludin function and is required for blood brain barrier maintenance: relevance to multiple sclerosis. 3. *J Neurosci* 2007 August 22;27(34): pp. 9032–9042.

[234] Michel CC, Neal CR. Openings through endothelial cells associated with increased microvascular permeability. 2. *Microcirculation* 1999 March;6(1): pp. 45–54.

[235] Michel CC, Curry FR. Glycocalyx volume: a critical review of tracer dilution methods for its measurement. 1. *Microcirculation* 2009 April;16(3): pp. 213–219.

[236]   Renkin EM. Relation of capillary morphology to transport of fluid and large molecules: a review. 40. *Acta Physiol Scand Suppl* 1979;463: pp. 81–91.

[237]   Stan RV, Kubitza M, Palade GE. PV-1 is a component of the fenestral and stomatal diaphragms in fenestrated endothelia. 2. *Proc Natl Acad Sci U S A* 1999 November 9;96(23): pp. 13203–13207.

[238]   Ioannidou S, Deinhardt K, Miotla J, Bradley J, Cheung E, Samuelsson S, Ng YS, Shima DT. An *in vitro* assay reveals a role for the diaphragm protein PV-1 in endothelial fenestra morphogenesis. 1. *Proc Natl Acad Sci U S A* 2006 November 7;103(45): pp. 16770–16775.

[239]   Esser S, Wolburg K, Wolburg H, Breier G, Kurzchalia T, Risau W. Vascular endothelial growth factor induces endothelial fenestrations *in vitro*. 3. *J Cell Biol* 1998 February 23;140(4): pp. 947–959.

[240]   Curry FE, Adamson RH. Transendothelial pathways in venular microvessels exposed to agents which increase permeability: the gaps in our knowledge. 14. *Microcirculation* 1999 March;6(1): pp. 3–5.

[241]   Malik AB, Lo SK. Vascular endothelial adhesion molecules and tissue inflammation. 3. *Pharmacol Rev* 1996 June;48(2): pp. 213–229.

[242]   McDonald DM, Thurston G, Baluk P. Endothelial gaps as sites for plasma leakage in inflammation. 34. *Microcirculation* 1999 March;6(1): pp. 7–22.

[243]   Kvietys PR, Sandig M. Neutrophil diapedesis: paracellular or transcellular? 2. *News Physiol Sci* 2001 February;16: pp. 15–19.

[244]   Muthuchamy M, Zawieja D. Molecular regulation of lymphatic contractility. 4. *Ann N Y Acad Sci* 2008;1131: pp. 89–99.

[245]   Gashev AA. Lymphatic vessels: pressure- and flow-dependent regulatory reactions. 7. *Ann N Y Acad Sci* 2008;1131: pp. 100–109.

[246]   Granger DN. Intestinal microcirculation and transmucosal fluid transport. *Am J Physiol* 1981 May;240(5): pp. G343–G349.

[247]   Kvietys PR, Patterson WG, Russell JM, Barrowman JA, Granger DN. Role of the microcirculation in ethanol-induced mucosal injury in the dog. 2. *Gastroenterology* 1984 September;87(3): pp. 562–571.

[248]   Granger DN, Barrowman JA, Harper SL, Kvietys PR, Korthuis RJ. Sympathetic stimulation and intestinal capillary fluid exchange. 10. *Am J Physiol* 1984 September;247(3 Pt 1): pp. G279–G283.

[249]   Granger DN, Mortillaro NA, Kvietys PR, Rutili G, Parker JC, Taylor AE. Role of the interstitial matrix during intestinal volume absorption. 4. *Am J Physiol* 1980 March;238(3): pp. G183–G189.

[250] Mortillaro NA, Taylor AE. Interaction of capillary and tissue forces in the cat small intestine. 12. *Circ Res* 1976 September;39(3): pp. 348–358.

[251] Barrowman JA, Granger DN. Effects of experimental cirrhosis on splanchnic microvascular fluid and solute exchange in the rat. 10. *Gastroenterology* 1984 July;87(1): pp. 165–172.

[252] Bohlen HG, Gore RW. Comparison of microvascular pressures and diameters in the innervated and denervated rat intestine. 4. *Microvasc Res* 1977 November;14(3): pp. 251–264.

[253] Mortillaro NA, Taylor AE. Interstitial fluid pressure of ileum measured from chronically implanted polyethylene capsules. 3. *Am J Physiol* 1989 July;257(1 Pt 2): pp. H62–H69.

[254] Lee JS. Lymph capillary pressure of rat intestinal villi during fluid absorption. *Am J Physiol* 1979;237: pp. E301–E307.

[255] Lee JS. Epithelial cell extrusion during fluid transport in canine small intestine. *Am J Physiol* 1977;232: pp. E408–E414.

[256] Altamirano M, Requena M, Perez TC. Interstitial fluid pressure in canine gastric mucosa. 1. *Am J Physiol* 1975 November;229(5): pp. 1414–1420.

[257] Barrowman JA, Perry MA, Kvietys PR, Ulrich M, Granger DN. Effects of bradykinin on intestinal transcapillary fluid exchange. 2. *Can J Physiol Pharmacol* 1981 August;59(8): pp. 786–789.

[258] Granger DN, Kvietys PR, Wilborn WH, Mortillaro NA, Taylor AE. Mechanism of glucagon-induced intestinal secretion. 4. *Am J Physiol* 1980 July;239(1): pp. G30–G38.

[259] Granger DN, Sennett M, McElearney P, Taylor AE. Effect of local arterial hypotension on cat intestinal capillary permeability. 1. *Gastroenterology* 1980 September;79(3): pp. 474–480.

[260] Lee J. Lymph pressure in intestinal villi and lymph flow during fluid secretion. In: *Tissue Fluid Pressure and Composition*. Hargens AR, ed. 165–172. 1981. Baltimore, MD: Williams & Wilkins.

[261] Altamirano M, Requena, Perez TC. Interstitial fluid pressure and alkaline gastric secretion. 2. *Am J Physiol* 1975 November;229(5): pp. 1421–1426.

[262] Mortillaro NA, Taylor AE. Microvascular permeability to endogenous plasma proteins in the jejunum. 2. *Am J Physiol* 1990 June;258(6 Pt 2): pp. H1650–H1654.

[263] Yablonski ME, Lifson N. Mechanism of production of intestinal secretion by elevated venous pressure. 2. *J Clin Invest* 1976 April;57(4): pp. 904–915.

[264] Johnson PC. Effect of venous pressure on mean capillary pressure and vascular resistance in the intestine. *Circ Res* 1965 March;16: pp. 294–300.

[265] Taylor AE. Capillary fluid filtration. Starling forces and lymph flow. *Circ Res* 1981 September;49(3): pp. 557–575.

[266] DiBona DR, Chen LC, Sharp GW. A study of intercellular spaces in the rabbit jejunum during acute volume expansion and after treatment with cholera toxin. 1. *J Clin Invest* 1974 May;53(5): pp. 1300–1307.

[267] Granger DN, Shackleford JS, Taylor AE. PGE1-induced intestinal secretion: mechanism of enhanced transmucosal protein efflux. 1. *Am J Physiol* 1979 June;236(6): pp. E788–E796.

[268] Granger DN, Mortillaro NA, Taylor AE. Interactions of intestinal lymph flow and secretion. 8. *Am J Physiol* 1977 January;232(1): pp. E13–E18.

[269] Mangino MJ, Chou CC. Thromboxane synthesis inhibitors and postprandial jejunal capillary exchange capacity. 2. *Am J Physiol* 1988 May;254(5 Pt 1): pp. G695–G701.

[270] Barrowman JA. *Physiology of the Gastrointestinal Lymphatic System*. 1978. Cambridge: Cambridge University Press.

[271] Borgstrom B, Laurell CB. Studies of lymph and lymph-proteins during absorption of fat and saline by rats. 1. *Acta Physiol Scand* 1953 October 6;29(2–3): pp. 264–280.

[272] Fraser JR, Gibson PR. Mechanisms by which food intake elevates circulating levels of hyaluronan in humans. 1. *J Intern Med* 2005 November;258(5): pp. 460–466.

[273] Granger HJ. Role of the interstitial matrix and lymphatic pump in regulation of transcapillary fluid balance. 99. *Microvasc Res* 1979 September;18(2): pp. 209–216.

[274] Bohlen HG, Unthank JL. Rat intestinal lymph osmolarity during glucose and oleic acid absorption. 6. *Am J Physiol* 1989 September;257(3 Pt 1): pp. G438–G446.

[275] Womack WA, Barrowman JA, Graham WH, Benoit JN, Kvietys PR, Granger DN. Quantitative assessment of villous motility. 1. *Am J Physiol* 1987 February;252(2 Pt 1): pp. G250–G256.

[276] Granger D, Barrowman J, Kvietys P. *Clinical Gastrointestinal Physiology*. 1985. Philadelphia, PA: Saunders.

[277] Sabesin SM, Frase S. Electron microscopic studies of the assembly, intracellular transport, and secretion of chylomicrons by rat intestine. 3. *J Lipid Res* 1977 July;18(4): pp. 496–511.

[278] Casley-Smith JR. The identification of chylomicra and lipoproteins in tissue sections and their passage into jejunal lacteals. *J Cell Biol* 1962 November;15: pp. 259–277.

[279] Dobbins WO, III, Rollins EL. Intestinal mucosal lymphatic permeability: an electron microscopic study of endothelial vesicles and cell junctions. 2. *J Ultrastruct Res* 1970 October;33(1): pp. 29–59.

[280] Tso P, Pitts V, Granger DN. Role of lymph flow in intestinal chylomicron transport. 1. *Am J Physiol* 1985 July;249(1 Pt 1): pp. G21–G28.

[281] Lee JS, Silverberg JW. Effect of cholera toxin on fluid absorption and villus lymph pressure in dog jejunal mucosa. 2. *Gastroenterology* 1972 May;62(5): pp. 993–1000.

[282] Cedgard S, Hallback DA, Jodal M, Lundgren O, Redfors S. The effects of cholera toxin on intramural blood flow distribution and capillary hydraulic conductivity in the cat small intestine. 1. *Acta Physiol Scand* 1978 February;102(2): pp. 148–158.

[283] Benoit JN, Navia CA, Taylor AE, Granger DN. Mathematical model of intestinal transcapillary fluid and protein exchange. In: *Physiology of the Intestinal Circulation.* Shepherd AP, Granger DN, eds. pp. 275–287. 1984. New York: Raven Press.

[284] Holzer P. Role of visceral afferent neurons in mucosal inflammation and defense. 18. *Curr Opin Pharmacol* 2007 December;7(6): pp. 563–569.

[285] Holzer P. Taste receptors in the gastrointestinal tract. V. Acid sensing in the gastrointestinal tract. 29. *Am J Physiol Gastrointest Liver Physiol* 2007 March;292(3): pp. G699–G705.

[286] Holzer P. Efferent-like roles of afferent neurons in the gut: blood flow regulation and tissue protection. 33. *Auton Neurosci* 2006 April 30;125(1–2): pp. 70–75.

[287] Wallace JL. Prostaglandins, NSAIDs, and gastric mucosal protection: why doesn't the stomach digest itself? 8. *Physiol Rev* 2008 October;88(4): pp. 1547–1565.

[288] Ham M, Kaunitz JD. Gastroduodenal defense. 5. *Curr Opin Gastroenterol* 2007 November;23(6): pp. 607–616.

[289] Akiba Y, Ghayouri S, Takeuchi T, Mizumori M, Guth PH, Engel E, Swenson ER, Kaunitz JD. Carbonic anhydrases and mucosal vanilloid receptors help mediate the hyperemic response to luminal CO_2 in rat duodenum. 13. *Gastroenterology* 2006 July;131(1): pp. 142–152.

[290] Shorrock CJ, Rees WD. Overview of gastroduodenal mucosal protection. 10. *Am J Med* 1988 February 22;84(2A): pp. 25–34.

[291] Wallace JL, Granger DN. The cellular and molecular basis of gastric mucosal defense. 2. *FASEB J* 1996 May;10(7): pp. 731–740.

[292] Clapham DE. TRP channels as cellular sensors. *Nature* 2003 December 4;426(6966): pp. 517–524.

[293] Kaunitz JD, Akiba Y. Luminal acid elicits a protective duodenal mucosal response. 22. *Keio J Med* 2002 March;51(1): pp. 29–35.

[294] Akiba Y, Nakamura M, Nagata H, Kaunitz JD, Ishii H. Acid-sensing pathways in rat gastrointestinal mucosa. 1. *J Gastroenterol* 2002 November;37 Suppl 14: pp. 133–138.

[295] Takeuchi K, Magee D, Critchlow J, Matthews J, Silen W. Studies of the pH gradient and thickness of frog gastric mucus gel. 2. *Gastroenterology* 1983 February;84(2): pp. 331–340.

[296] Engel E, Guth PH, Nishizaki Y, Kaunitz JD. Barrier function of the gastric mucus gel. 21. *Am J Physiol* 1995 December;269(6 Pt 1): pp. G994–G999.

[297] Starlinger M, Schiessel R, Hung CR, Silen W. H+ back diffusion stimulating gastric mucosal blood flow in the rabbit fundus. 1. *Surgery* 1981 February;89(2): pp. 232–236.

[298]   Abdel-Salam OM, Czimmer J, Debreceni A, Szolcsanyi J, Mozsik G. Gastric mucosal integrity: gastric mucosal blood flow and microcirculation. An overview. 1. *J Physiol Paris* 2001 January;95(1–6): pp. 105–127.

[299]   Wallace JL, McKnight GW. The mucoid cap over superficial gastric damage in the rat. A high-pH microenvironment dissipated by nonsteroidal antiinflammatory drugs and endothelin. 12. *Gastroenterology* 1990 August;99(2): pp. 295–304.

[300]   Merchant NB, Dempsey DT, Grabowski MW, Rizzo M, Ritchie WP, Jr. Capsaicin-induced gastric mucosal hyperemia and protection: the role of calcitonin gene-related peptide. 3. *Surgery* 1994 August;116(2): pp. 419–425.

[301]   Podolsky RS, Grabowski M, Milner R, Ritchie WP, Dempsey DT. Capsaicin-induced gastric hyperemia and protection are NO-dependent. 1. *J Surg Res* 1994 October;57(4): pp. 438–442.

[302]   Sullivan TR, Jr, Milner R, Dempsey DT, Ritchie WP, Jr. Effect of capsaicin on gastric mucosal injury and blood flow following bile acid exposure. 2. *J Surg Res* 1992 June;52(6): pp. 596–600.

[303]   Sullivan TR, Jr, Dempsey DT, Milner R, Ritchie WP, Jr. Effect of local acid–base status on gastric mucosal blood flow and surface cell injury by bile acid. 1. *J Surg Res* 1994 January;56(1): pp. 112–116.

[304]   Hofmann AF, Borgstrom B. The intraluminal ohase of fat digestion in man: the lipid content of the micellar and oil phases of intestinal content obtained during fat digestion and absorption. *J Clin Invest* 1964;43: pp. 247–257.

[305]   Kvietys PR, Specian RD, Grisham MB, Tso P. Jejunal mucosal injury and restitution: role of hydrolytic products of food digestion. 4. *Am J Physiol* 1991 September;261(3 Pt 1): pp. G384–G391.

[306]   Velasquez OR, Henninger K, Fowler M, Tso P, Crissinger KD. Oleic acid-induced mucosal injury in developing piglet intestine. 1. *Am J Physiol* 1993 March;264(3 Pt 1): pp. G576–G582.

[307]   Velasquez OR, Place AR, Tso P, Crissinger KD. Developing intestine is injured during absorption of oleic acid but not its ethyl ester. 1. *J Clin Invest* 1994 February;93(2): pp. 479–485.

[308]   Cepinskas G, Specian RD, Kvietys PR. Adaptive cytoprotection in the small intestine: role of mucus. 2. *Am J Physiol* 1993 May;264(5 Pt 1): pp. G921–G927.

[309]   Ammon HV, Thomas PJ, Phillips SF. Effects of oleic and ricinoleic acids on net jejunal water and electrolyte movement. Perfusion studies in man. 3. *J Clin Invest* 1974 February;53(2): pp. 374–379.

[310] Ammon HV, Thomas PJ, Phillips SF. Effects of long chain fatty acids on solute absorption: perfusion studies in the human jejunum. 2. *Gut* 1977 October;18(10): pp. 805–813.

[311] Kvietys PR, Wilborn WH, Granger DN. Effect of atropine on bile–oleic acid-induced alterations in dog jejunal hemodynamics, oxygenation, and net transmucosal water movement. 2. *Gastroenterology* 1981 January;80(1): pp. 31–38.

[312] Lapre JA, Termont DS, Groen AK, Van der MR. Lytic effects of mixed micelles of fatty acids and bile acids. 3. *Am J Physiol* 1992 September;263(3 Pt 1): pp. G333–G337.

[313] Pawlik WW, Gustaw P, Jacobson ED, Sendur R, Czarnobilski K. Nitric oxide mediates intestinal hyperaemic responses to intraluminal bile-oleate. 2. *Pflugers Arch* 1995 January;429(3): pp. 301–305.

[314] Charman WN, Porter CJ, Mithani S, Dressman JB. Physiochemical and physiological mechanisms for the effects of food on drug absorption: the role of lipids and pH. 1. *J Pharm Sci* 1997 March;86(3): pp. 269–282.

[315] Pingle SC, Matta JA, Ahern GP. Capsaicin receptor: TRPV1 a promiscuous TRP channel. 1. *Handb Exp Pharmacol* 2007;(179): pp. 155–171.

[316] Thabuis C, Tissot-Favre D, Bezelgues JB, Martin JC, Cruz-Hernandez C, Dionisi F, Destaillats F. Biological functions and metabolism of oleoylethanolamide. 1. *Lipids* 2008 October;43(10): pp. 887–894.

[317] Schwartz GJ, Fu J, Astarita G, Li X, Gaetani S, Campolongo P, Cuomo V, Piomelli D. The lipid messenger OEA links dietary fat intake to satiety. 1. *Cell Metab* 2008 October;8(4): pp. 281–288.

[318] Ahern GP. Activation of TRPV1 by the satiety factor oleoylethanolamide. 15. *J Biol Chem* 2003 August 15;278(33): pp. 30429–30434.

[319] Hansen MB, Witte AB. The role of serotonin in intestinal luminal sensing and secretion. 1. *Acta Physiol (Oxf)* 2008 August;193(4): pp. 311–323.

[320] Savastano DM, Hayes MR, Covasa M. Serotonin-type 3 receptors mediate intestinal lipid-induced satiation and Fos-like immunoreactivity in the dorsal hindbrain. 1. *Am J Physiol Regul Integr Comp Physiol* 2007 March;292(3): pp. R1063–R1070.

[321] Ishikawa S, Cepinskas G, Specian RD, Itoh M, Kvietys PR. Epidermal growth factor attenuates jejunal mucosal injury induced by oleic acid: role of mucus. 1. *Am J Physiol* 1994 December;267(6 Pt 1): pp. G1067–G1077.

[322] Argenzio RA, Meuten DJ. Short-chain fatty acids induce reversible injury of porcine colon. 1. *Dig Dis Sci* 1991 October;36(10): pp. 1459–1468.

[323] Topping DL, Clifton PM. Short-chain fatty acids and human colonic function: roles of resistant starch and nonstarch polysaccharides. 1. *Physiol Rev* 2001 July;81(3): pp. 1031–1064.

[324] Nafday SM, Chen W, Peng L, Babyatsky MW, Holzman IR, Lin J. Short-chain fatty acids induce colonic mucosal injury in rats with various postnatal ages. 3. *Pediatr Res* 2005 February;57(2): pp. 201–204.

[325] Flourie B, Florent C, Jouany JP, Thivend P, Etanchaud F, Rambaud JC. Colonic metabolism of wheat starch in healthy humans. Effects on fecal outputs and clinical symptoms. 14. *Gastroenterology* 1986 January;90(1): pp. 111–119.

[326] Barcelo A, Claustre J, Moro F, Chayvialle JA, Cuber JC, Plaisancie P. Mucin secretion is modulated by luminal factors in the isolated vascularly perfused rat colon. 3. *Gut* 2000 February;46(2): pp. 218–224.

[327] Fukumoto S, Tatewaki M, Yamada T, Fujimiya M, Mantyh C, Voss M, Eubanks S, Harris M, Pappas TN, Takahashi T. Short-chain fatty acids stimulate colonic transit via intraluminal 5-HT release in rats. 1. *Am J Physiol Regul Integr Comp Physiol* 2003 May;284(5): pp. R1269–R1276.

[328] Paimela H, Goddard PJ, Carter K, Khakee R, McNeil PL, Ito S, Silen W. Restitution of frog gastric mucosa in vitro: effect of basic fibroblast growth factor. 3. *Gastroenterology* 1993 May;104(5): pp. 1337–1345.

[329] Paimela H, Goddard PJ, Silen W. Present views on restitution of gastrointestinal epithelium. 1. *Dig Dis Sci* 1995 November;40(11): pp. 2495–2496.

[330] Mammen JM, Matthews JB. Mucosal repair in the gastrointestinal tract. 3. *Crit Care Med* 2003 August;31(8 Suppl): pp. S532–S537.

[331] Silen W, Ito S. Mechanisms for rapid re-epithelialization of the gastric mucosal surface. 11. *Annu Rev Physiol* 1985;47: pp. 217–229.

[332] Blikslager AT, Moeser AJ, Gookin JL, Jones SL, Odle J. Restoration of barrier function in injured intestinal mucosa. 10. *Physiol Rev* 2007 April;87(2): pp. 545–564.

[333] Sturm A, Dignass AU. Epithelial restitution and wound healing in inflammatory bowel disease. 4. *World J Gastroenterol* 2008 January 21;14(3): pp. 348–353.

[334] Lacy ER. Rapid epithelial restitution in the stomach: an updated perspective. 22. *Scand J Gastroenterol Suppl* 1995;210: pp. 6–8.

[335] Evangelista S. Role of calcitonin gene-related peptide in gastric mucosal defence and healing. 2. *Curr Pharm Des* 2009;15(30): pp. 3571–3576.

[336] Wilson AJ, Gibson PR. Epithelial migration in the colon: filling in the gaps. 6. *Clin Sci (Lond)* 1997 August;93(2): pp. 97–108.

[337] Grant R, Grossman MI, Ivy AC. Histological changes in the gastric mucosa during digestion and their relationship to mucosal growth. 1. *Gastroenterology* 1953 October;25(2): pp. 218–231.

[338]   Premen AJ, Banchs V, Womack WA, Kvietys PR, Granger DN. Importance of collateral circulation in the vascularly occluded feline intestine. 1. *Gastroenterology* 1987 May;92(5 Pt 1): pp. 1215–1219.

[339]   Bulkley GB, Womack WA, Downey JM, Kvietys PR, Granger DN. Characterization of segmental collateral blood flow in the small intestine. 2. *Am J Physiol* 1985 August;249(2 Pt 1): pp. G228–G235.

[340]   Perry MA, Haedicke GJ, Bulkley GB, Kvietys PR, Granger DN. Relationship between acid secretion and blood flow in the canine stomach: role of oxygen consumption. 1. *Gastroenterology* 1983 September;85(3): pp. 529–534.

[341]   Varro V, Csernay L, Szarvas F, Blaho G. Effect of glucose and glycine solution on the circulation of the isolated jejunal loop in the dog. 4. *Am J Dig Dis* 1967 January;12(1): pp. 60–64.

[342]   Walus KM, Jacobson ED. Relation between small intestinal motility and circulation. 3. *Am J Physiol* 1981 July;241(1): pp. G1–15.

[343]   Womack WA, Tygart PK, Mailman D, Kvietys PR, Granger DN. Villous motility: relationship to lymph flow and blood flow in the dog jejunum. 1. *Gastroenterology* 1988 April;94(4): pp. 977–983.

[344]   Moses FM. Exercise-associated intestinal ischemia. 3. *Curr Sports Med Rep* 2005 April;4(2): pp. 91–95.

[345]   Oldenburg WA, Lau LL, Rodenberg TJ, Edmonds HJ, Burger CD. Acute mesenteric ischemia: a clinical review. 5. *Arch Intern Med* 2004 May 24;164(10): pp. 1054–1062.

[346]   Carden DL, Granger DN. Pathophysiology of ischaemia–reperfusion injury. 1. *J Pathol* 2000 February;190(3): pp. 255–266.

[347]   Leung FW, Su KC, Passaro E Jr, Guth PH. Regional differences in gut blood flow and mucosal damage in response to ischemia and reperfusion. 3. *Am J Physiol* 1992 September;263(3 Pt 1): pp. G301–G305.

[348]   Takeyoshi I, Zhang S, Nakamura K, Ikoma A, Zhu Y, Starzl TE, Todo S. Effect of ischemia on the canine large bowel: a comparison with the small intestine. 4. *J Surg Res* 1996 April;62(1): pp. 41–48.

[349]   Fukuyama K, Iwakiri R, Noda T, Kojima M, Utsumi H, Tsunada S, Sakata H, Ootani A, Fujimoto K. Apoptosis induced by ischemia–reperfusion and fasting in gastric mucosa compared to small intestinal mucosa in rats. 1. *Dig Dis Sci* 2001 March;46(3): pp. 545–549.

[350]   Jodal M, Lundgren O. Countercurrent mechanisms in the mammalian gastrointestinal tract. 39. *Gastroenterology* 1986 July;91(1): pp. 225–241.

[351]   Levitt DG, Bond JH, Levitt MD. Use of a model of small bowel mucosa to predict passive absorption. 3. *Am J Physiol* 1980 July;239(1): pp. G23–G29.

[352] Taylor CT, Colgan SP. Hypoxia and gastrointestinal disease. 1. *J Mol Med* 2007 December;85(12): pp. 1295–1300.

[353] Haglund U. Gut ischaemia. *Gut* 1994 January;35(1 Suppl): pp. S73–S76.

[354] Parks DA, Grogaard B, Granger DN. Comparison of partial and complete arterial occlusion models for studying intestinal ischemia. 1. *Surgery* 1982 November;92(5): pp. 896–901.

[355] Chiu CJ, McArdle AH, Brown R, Scott HJ, Gurd FN. Intestinal mucosal lesion in low-flow states. I. A morphological, hemodynamic, and metabolic reappraisal. 5. *Arch Surg* 1970 October;101(4): pp. 478–483.

[356] Derikx JP, Matthijsen RA, de Bruine AP, van Bijnen AA, Heineman E, van Dam RM, Dejong CH, Buurman WA. Rapid reversal of human intestinal ischemia–reperfusion induced damage by shedding of injured enterocytes and reepithelialisation. 10. *PLoS One* 2008;3(10): p. e3428.

[357] Matthijsen RA, Derikx JP, Kuipers D, van Dam RM, Dejong CH, Buurman WA. Enterocyte shedding and epithelial lining repair following ischemia of the human small intestine attenuate inflammation. 1. *PLoS One* 2009;4(9): p. e7045.

[358] Ahren C, Haglund U. Mucosal lesions in the small intestine of the cat during low flow. 4. *Acta Physiol Scand* 1973 August;88(4): pp. 541–550.

[359] Bulkley GB. Free radical-mediated reperfusion injury: a selective review. 94. *Br J Cancer Suppl* 1987 June;8: pp. 66–73.

[360] Bulkley GB, Kvietys PR, Parks DA, Perry MA, Granger DN. Relationship of blood flow and oxygen consumption to ischemic injury in the canine small intestine. 2. *Gastroenterology* 1985 October;89(4): pp. 852–857.

[361] Perry MA, Wadhwa SS. Gradual reintroduction of oxygen reduces reperfusion injury in cat stomach. 1. *Am J Physiol* 1988 March;254(3 Pt 1): pp. G366–G372.

[362] Robinson JW, Haroud M, Winistorfer B, Mirkovitch V. Recovery of function and structure of dog ileum and colon following two hours' acute ischaemia. 1. *Eur J Clin Invest* 1974 December 5;4(6): pp. 443–452.

[363] Granger DN, Rutili G, McCord JM. Superoxide radicals in feline intestinal ischemia. 1. *Gastroenterology* 1981 July;81(1): pp. 22–29.

[364] Osborne DL, Aw TY, Cepinskas G, Kvietys PR. Development of ischemia/reperfusion tolerance in the rat small intestine. An epithelium-independent event. 1. *J Clin Invest* 1994 November;94(5): pp. 1910–1918.

[365] Kanwar S, Hickey MJ, Kubes P. Postischemic inflammation: a role for mast cells in intestine but not in skeletal muscle. 3. *Am J Physiol* 1998 August;275(2 Pt 1): pp. G212–G218.

[366] Granger DN. Role of xanthine oxidase and granulocytes in ischemia–reperfusion injury. *Am J Physiol* 1988 December;255(6 Pt 2): pp. H1269–H1275.

[367]   Granger DN. Ischemia–reperfusion: mechanisms of microvascular dysfunction and the influence of risk factors for cardiovascular disease. *Microcirculation* 1999 September;6(3): pp. 167–178.

[368]   Massberg S, Messmer K. The nature of ischemia/reperfusion injury. 7. *Transplant Proc* 1998 December;30(8): pp. 4217–4223.

[369]   Kvietys PR, Granger DN. Endothelial cell monolayers as a tool for studying microvascular pathophysiology. 2. *Am J Physiol* 1997 December;273(6 Pt 1): pp. G1189–G1199.

[370]   Jarasch ED, Bruder G, Heid HW. Significance of xanthine oxidase in capillary endothelial cells. 1. *Acta Physiol Scand Suppl* 1986;548: pp. 39–46.

[371]   Ratych RE, Chuknyiska RS, Bulkley GB. The primary localization of free radical generation after anoxia/reoxygenation in isolated endothelial cells. 1. *Surgery* 1987 August;102(2): pp. 122–131.

[372]   Grisham MB, Hernandez LA, Granger DN. Xanthine oxidase and neutrophil infiltration in intestinal ischemia. *Am J Physiol* 1986 October;251(4 Pt 1): pp. G567–G574.

[373]   Hernandez LA, Grisham MB, Twohig B, Arfors KE, Harlan JM, Granger DN. Role of neutrophils in ischemia–reperfusion-induced microvascular injury. 1. *Am J Physiol* 1987 September;253(3 Pt 2): pp. H699–H703.

[374]   Yoshida N, Granger DN, Anderson DC, Rothlein R, Lane C, Kvietys PR. Anoxia/reoxygenation-induced neutrophil adherence to cultured endothelial cells. 1. *Am J Physiol* 1992 June;262(6 Pt 2): pp. H1891–H1898.

[375]   Heller T, Hennecke M, Baumann U, Gessner JE, zu Vilsendorf AM, Baensch M, Boulay F, Kola A, Klos A, Bautsch W, Kohl J. Selection of a C5a receptor antagonist from phage libraries attenuating the inflammatory response in immune complex disease and ischemia/reperfusion injury. 2. *J Immunol* 1999 July 15;163(2): pp. 985–994.

[376]   Ichikawa H, Flores S, Kvietys PR, Wolf RE, Yoshikawa T, Granger DN, Aw TY. Molecular mechanisms of anoxia/reoxygenation-induced neutrophil adherence to cultured endothelial cells. 1. *Circ Res* 1997 December;81(6): pp. 922–931.

[377]   Grisham MB, Granger DN, Lefer DJ. Modulation of leukocyte–endothelial interactions by reactive metabolites of oxygen and nitrogen: relevance to ischemic heart disease. 7. *Free Radic Biol Med* 1998 September;25(4–5): pp. 404–433.

[378]   Kubes P, Suzuki M, Granger DN. Nitric oxide: an endogenous modulator of leukocyte adhesion. 2. *Proc Natl Acad Sci U S A* 1991 June 1;88(11): pp. 4651–4655.

[379]   Land W, Messmer K. The impact of ischemia/reperfusion injury on specific and non-specific, early and late chronic events after organ transplantation. *Transplant Rev* 1996;10: pp. 108–127.

[380] Mallick IH, Yang W, Winslet MC, Seifalian AM. Ischemia–reperfusion injury of the intestine and protective strategies against injury. 8. *Dig Dis Sci* 2004 September;49(9): pp. 1359–1377.

[381] Granger D, Kevil C, Grisham M. Recruitment of inflammatory and immune cells in the gut: physiology and pathophysiology.

[382] Wang Z, Rui T, Yang M, Valiyeva F, Kvietys PR. Alveolar macrophages from septic mice promote polymorphonuclear leukocyte transendothelial migration via an endothelial cell Src kinase/NADPH oxidase pathway. 1. *J Immunol* 2008 December 15;181(12): pp. 8735–8744.

[383] Tailor A, Cooper D, Granger DN. Platelet–vessel wall interactions in the microcirculation. 1. *Microcirculation* 2005 April;12(3): pp. 275–285.

[384] Massberg S, Enders G, Leiderer R, Eisenmenger S, Vestweber D, Krombach F, Messmer K. Platelet–endothelial cell interactions during ischemia/reperfusion: the role of P-selectin. 4. *Blood* 1998 July 15;92(2): pp. 507–515.

[385] Cooper D, Russell J, Chitman KD, Williams MC, Wolf RE, Granger DN. Leukocyte dependence of platelet adhesion in postcapillary venules. 1. *Am J Physiol Heart Circ Physiol* 2004 May;286(5): pp. H1895–H1900.

[386] Linfert D, Chowdhry T, Rabb H. Lymphocytes and ischemia–reperfusion injury. 1. *Transplant Rev (Orlando)* 2009 January;23(1): pp. 1–10.

[387] Shigematsu T, Wolf RE, Granger DN. T-lymphocytes modulate the microvascular and inflammatory responses to intestinal ischemia–reperfusion. 1. *Microcirculation* 2002 April;9(2): pp. 99–109.

[388] Kokura S, Wolf RE, Yoshikawa T, Ichikawa H, Granger DN, Aw TY. Endothelial cells exposed to anoxia/reoxygenation are hyperadhesive to T-lymphocytes: kinetics and molecular mechanisms. 3. *Microcirculation* 2000 February;7(1): pp. 13–23.

[389] Kokura S, Wolf RE, Yoshikawa T, Granger DN, Aw TY. T-lymphocyte-derived tumor necrosis factor exacerbates anoxia-reoxygenation-induced neutrophil–endothelial cell adhesion. 4. *Circ Res* 2000 February 4;86(2): pp. 205–213.

[390] Osman M, Russell J, Granger DN. Lymphocyte-derived interferon-gamma mediates ischemia–reperfusion-induced leukocyte and platelet adhesion in intestinal microcirculation. 1. *Am J Physiol Gastrointest Liver Physiol* 2009 March;296(3): pp. G659–G663.

[391] Brzozowski T, Konturek PC, Konturek SJ, Drozdowicz D, Kwiecien S, Pajdo R, Bielanski W, Hahn EG. Role of gastric acid secretion in progression of acute gastric erosions induced by ischemia–reperfusion into gastric ulcers. 46. *Eur J Pharmacol* 2000 June 9;398(1): pp. 147–158.

[392] Kotani T, Murashima Y, Kobata A, Amagase K, Takeuchi K. Pathogenic importance of pepsin in ischemia/reperfusion-induced gastric injury. 2. *Life Sci* 2007 May 1;80(21): pp. 1984–1992.

[393] Montgomery A, Borgstrom A, Haglund U. Pancreatic proteases and intestinal mucosal injury after ischemia and reperfusion in the pig. 3. *Gastroenterology* 1992 January;102(1): pp. 216–222.

[394] Mitsuoka H, Kistler EB, Schmid-Schonbein GW. Generation of *in vivo* activating factors in the ischemic intestine by pancreatic enzymes. 3. *Proc Natl Acad Sci U S A* 2000 February 15;97(4): pp. 1772–1777.

[395] Langkamp-Henken B, Kudsk KA, Proctor KG. Fasting-induced reduction of intestinal reperfusion injury. 3. *JPEN J Parenter Enteral Nutr* 1995 March;19(2): pp. 127–132.

[396] Crissinger KD, Tso P. The role of lipids in ischemia/reperfusion-induced changes in mucosal permeability in developing piglets. 4. *Gastroenterology* 1992 May;102(5): pp. 1693–1699.

[397] Schmid-Schonbein GW. Biomechanical aspects of the auto-digestion theory. 9. *Mol Cell Biomech* 2008 June;5(2): pp. 83–95.

[398] Penn AH, Schmid-Schonbein GW. The intestine as source of cytotoxic mediators in shock: free fatty acids and degradation of lipid-binding proteins. 1. *Am J Physiol Heart Circ Physiol* 2008 April;294(4): pp. H1779–H1792.

[399] Cepinskas G, Rui T, Kvietys PR. Interaction between reactive oxygen metabolites and nitric oxide in oxidant tolerance. 3. *Free Radic Biol Med* 2002 August 15;33(4): pp. 433–440.

[400] Lu D, Maulik N, Moraru II, Kreutzer DL, Das DK. Molecular adaptation of vascular endothelial cells to oxidative stress. 2. *Am J Physiol* 1993 March;264(3 Pt 1): pp. C715–C722.

[401] Cepinskas G, Lush CW, Kvietys PR. Anoxia/reoxygenation-induced tolerance with respect to polymorphonuclear leukocyte adhesion to cultured endothelial cells. A nuclear factor-κB-mediated phenomenon. 4. *Circ Res* 1999 January 8;84(1): pp. 103–112.

[402] Davis JM, Gute DC, Jones S, Krsmanovic A, Korthuis RJ. Ischemic preconditioning prevents postischemic P-selectin expression in the rat small intestine. 1. *Am J Physiol* 1999 December;277(6 Pt 2): pp. H2476–H2481.

[403] Park PO, Haglund U. Regeneration of small bowel mucosa after intestinal ischemia. 2. *Crit Care Med* 1992 January;20(1): pp. 135–139.

[404] Colle I, Geerts AM, Van SC, Van VH. Hemodynamic changes in splanchnic blood vessels in portal hypertension. 4. *Anat Rec (Hoboken)* 2008 June;291(6): pp. 699–713.

[405] Cichoz-Lach H, Celinski K, Slomka M, Kasztelan-Szczerbinska B. Pathophysiology of portal hypertension. 3. *J Physiol Pharmacol* 2008 August;59 Suppl 2: pp. 231–238.

[406] Bosch J, Berzigotti A, Garcia-Pagan JC, Abraldes JG. The management of portal hypertension: rational basis, available treatments and future options. 4. *J Hepatol* 2008;48 Suppl 1: pp. S68–S92.

[407] Benoit J, Granger DN. Chronic portal hypertension and the splanchnic circulation. In: *Pathophysiology of the Splanchnic Circulation* (Vol I). Kvietys PR, Barrowman JA, Granger DN, eds. pp. 57–88. 1987. Boca Raton, FL: CRC Press.

[408] Kashani A, Landaverde C, Medici V, Rossaro L. Fluid retention in cirrhosis: pathophysiology and management. 1. *QJM* 2008 February;101(2): pp. 71–85.

[409] Bosch J, Garcia-Pagan JC. Complications of cirrhosis. I. Portal hypertension. 92. *J Hepatol* 2000;32(1 Suppl): pp. 141–156.

[410] La VG, Gentilini P. Hemodynamic alterations in liver cirrhosis. 2. *Mol Aspects Med* 2008 February;29(1–2): pp. 112–118.

[411] Gatta A, Bolognesi M, Merkel C. Vasoactive factors and hemodynamic mechanisms in the pathophysiology of portal hypertension in cirrhosis. 4. *Mol Aspects Med* 2008 February;29(1–2): pp. 119–129.

[412] Aller MA, Nava MP, Cuellar C, Chivato T, Arias JL, Sanchez-Patan F, de VF, Alvarez E, Arias J. Evolutive phases of experimental prehepatic portal hypertension. 6. *J Gastroenterol Hepatol* 2007 July;22(7): pp. 1127–1133.

[413] Iwakiri Y, Groszmann RJ. The hyperdynamic circulation of chronic liver diseases: from the patient to the molecule. 7. *Hepatology* 2006 February;43(2 Suppl 1): pp. S121–S131.

[414] Geerts AM, Vanheule E, Praet M, Van VH, De VM, Colle I. Comparison of three research models of portal hypertension in mice: macroscopic, histological and portal pressure evaluation. 2. *Int J Exp Pathol* 2008 August;89(4): pp. 251–263.

[415] Sikuler E, Kravetz D, Groszmann RJ. Evolution of portal hypertension and mechanisms involved in its maintenance in a rat model. 2. *Am J Physiol* 1985 June;248(6 Pt 1): pp. G618–G625.

[416] Tsai MH, Iwakiri Y, Cadelina G, Sessa WC, Groszmann RJ. Mesenteric vasoconstriction triggers nitric oxide overproduction in the superior mesenteric artery of portal hypertensive rats. 1. *Gastroenterology* 2003 November;125(5): pp. 1452–1461.

[417] Abraldes JG, Iwakiri Y, Loureiro-Silva M, Haq O, Sessa WC, Groszmann RJ. Mild increases in portal pressure upregulate vascular endothelial growth factor and endothelial nitric oxide synthase in the intestinal microcirculatory bed, leading to a hyperdynamic state. 2. *Am J Physiol Gastrointest Liver Physiol* 2006 May;290(5): pp. G980–G987.

[418] Benoit JN, Barrowman JA, Harper SL, Kvietys PR, Granger DN. Role of humoral factors in the intestinal hyperemia associated with chronic portal hypertension. 5. *Am J Physiol* 1984 November;247(5 Pt 1): pp. G486–G493.

[419] Benoit JN, Womack WA, Korthuis RJ, Wilborn WH, Granger DN. Chronic portal hypertension: effects on gastrointestinal blood flow distribution. 2. *Am J Physiol* 1986 April;250 (4 Pt 1): pp. G535–G539.

[420] Benoit JN, Womack WA, Hernandez L, Granger DN. "Forward" and "backward" flow mechanisms of portal hypertension. Relative contributions in the rat model of portal vein stenosis. 3. *Gastroenterology* 1985 November;89(5): pp. 1092–1096.

[421] Fernandez M, Mejias M, Angermayr B, Garcia-Pagan JC, Rodes J, Bosch J. Inhibition of VEGF receptor-2 decreases the development of hyperdynamic splanchnic circulation and portal–systemic collateral vessels in portal hypertensive rats. 13. *J Hepatol* 2005 July;43(1): pp. 98–103.

[422] Fernandez M, Vizzutti F, Garcia-Pagan JC, Rodes J, Bosch J. Anti-VEGF receptor-2 monoclonal antibody prevents portal–systemic collateral vessel formation in portal hypertensive mice. 14. *Gastroenterology* 2004 March;126(3): pp. 886–894.

[423] Laragh JH, Cannon PJ, Bentzel CJ, Sicinski AM, Meltzer JI. Angiotensin II, norepinephrine, and renal transport of electrolytes and water in normal man and in cirrhosis with ascites. 1. *J Clin Invest* 1963 July;42(7): pp. 1179–1192.

[424] Finberg JP, Syrop HA, Better OS. Blunted pressor response to angiotensin and sympathomimetic amines in bile-duct ligated dogs. 1. *Clin Sci (Lond)* 1981 November;61(5): pp. 535–539.

[425] Murray BM, Paller MS. Decreased pressor reactivity to angiotensin II in cirrhotic rats. Evidence for a post-receptor defect in angiotensin action. 4. *Circ Res* 1985 September;57(3): pp. 424–431.

[426] Kitano S, Koyanagi N, Sugimachi K, Kobayashi M, Inokuchi K. Mucosal blood flow and modified vascular responses to norepinephrine in the stomach of rats with liver cirrhosis. 20. *Eur Surg Res* 1982;14(3): pp. 221–230.

[427] Kiel JW, Pitts V, Benoit JN, Granger DN, Shepherd AP. Reduced vascular sensitivity to norepinephrine in portal-hypertensive rats. 1. *Am J Physiol* 1985 February;248(2 Pt 1): pp. G192–G195.

[428] Benoit JN, Zimmerman B, Premen AJ, Go VL, Granger DN. Role of glucagon in splanchnic hyperemia of chronic portal hypertension. 15. *Am J Physiol* 1986 November;251(5 Pt 1): pp. G674–G677.

[429] Atucha NM, Shah V, Garcia-Cardena G, Sessa WE, Groszmann RJ. Role of endothelium in the abnormal response of mesenteric vessels in rats with portal hypertension and liver cirrhosis. 1. *Gastroenterology* 1996 December;111(6): pp. 1627–1632.

[430] Iwakiri Y, Cadelina G, Sessa WC, Groszmann RJ. Mice with targeted deletion of eNOS develop hyperdynamic circulation associated with portal hypertension. 8. *Am J Physiol Gastrointest Liver Physiol* 2002 November;283(5): pp. G1074–G1081.

[431] Theodorakis NG, Wang YN, Skill NJ, Metz MA, Cahill PA, Redmond EM, Sitzmann JV. The role of nitric oxide synthase isoforms in extrahepatic portal hypertension: studies in gene-knockout mice. 3. *Gastroenterology* 2003 May;124(5): pp. 1500–1508.

[432] Theodorakis NG, Wang YN, Wu JM, Maluccio MA, Sitzmann JV, Skill NJ. Role of endothelial nitric oxide synthase in the development of portal hypertension in the carbon tetrachloride induced liver fibrosis model. 1. *Am J Physiol Gastrointest Liver Physiol* 2009 July 23.

[433] Batkai S, Jarai Z, Wagner JA, Goparaju SK, Varga K, Liu J, Wang L, Mirshahi F, Khanolkar AD, Makriyannis A, Urbaschek R, Garcia N, Jr, Sanyal AJ, Kunos G. Endocannabinoids acting at vascular CB1 receptors mediate the vasodilated state in advanced liver cirrhosis. 5. *Nat Med* 2001 July;7(7): pp. 827–832.

[434] Moezi L, Gaskari SA, Lee SS. Endocannabinoids and liver disease. V. endocannabinoids as mediators of vascular and cardiac abnormalities in cirrhosis. 1. *Am J Physiol Gastrointest Liver Physiol* 2008 October;295(4): pp. G649–G653.

[435] Granger DN, Barrowman JA. Microcirculation of the alimentary tract. II. Pathophysiology of edema. *Gastroenterology* 1983 May;84(5 Pt 1): pp. 1035–1049.

[436] Norman DA, Atkins JM, Seelig LL, Jr, Gomez-Sanchez C, Krejs GJ. Water and electrolyte movement and mucosal morphology in the jejunum of patients with portal hypertension. 1. *Gastroenterology* 1980 October;79(4): pp. 707–715.

[437] Ohta M, Yamaguchi S, Gotoh N, Tomikawa M. Pathogenesis of portal hypertensive gastropathy: a clinical and experimental review. 1. *Surgery* 2002 January;131(1 Suppl): pp. S165–S170.

[438] Perini RF, Camara PR, Ferraz JG. Pathogenesis of portal hypertensive gastropathy: translating basic research into clinical practice. 1. *Nat Clin Pract Gastroenterol Hepatol* 2009 March;6(3): pp. 150–158.

[439] Tomikawa M, Akiba Y, Kaunitz JD, Kawanaka H, Sugimachi K, Sarfeh IJ, Tarnawski AS. New insights into impairment of mucosal defense in portal hypertensive gastric mucosa. 1. *J Gastrointest Surg* 2000 September;4(5): pp. 458–463.

[440] Nishizaki Y, Guth PH, Sternini C, Kaunitz JD. Impairment of the gastric hyperemic response to luminal acid in cirrhotic rats. 1. *Am J Physiol* 1996 January;270(1 Pt 1): pp. G71–G78.

[441] Iwao T, Toyonaga A, Ikegami M, Shigemori H, Oho K, Sumino M, Tanikawa K. Gastric mucus generation in cirrhotic patients with portal hypertension. Effects of tetraprenylacetone. 21. *Dig Dis Sci* 1996 September;41(9): pp. 1727–1732.

[442] Beck PL, McKnight W, Lee SS, Wallace JL. Prostaglandin modulation of the gastric vasculature and mucosal integrity in cirrhotic rats. 4. *Am J Physiol* 1993 September;265(3 Pt 1): pp. G453–G458.

Author Biography

Dr. Peter R. Kvietys is a scientist at the Lawson Health Research Institute at London Health Sciences Center (LHSC), and a Professor in the Department of Medicine at the University of Western Ontario London, Ontario, Canada. He received his Ph.D. from Michigan State University. He established and directed the Division of Vascular Biology at LHSC. He has chaired or co-chaired FASEB conferences on the topic of "Gastrointestinal Circulation" and served as an associate editor of the *American Journal of Physiology: Gastrointestinal & Liver Physiology.* Dr. Kvietys has been a consultant for various government agencies both in the U.S. and Canada. He has published more than 160 papers, 25 chapters in books, and 3 books on the gastrointestinal circulation and related topics.